THE FOOD AND BEVERAGE HOSPITALITY INDUSTRY IN INDIA

An Emergent Segment

THE FOOD AND BEVERAGE HOSPITALITY INDUSTRY IN INDIA

An Emergent Segment

Sandeep Munjal
Sanjay Sharma

AAP APPLE ACADEMIC PRESS

First edition published 2022

Apple Academic Press Inc.
1265 Goldenrod Circle, NE,
Palm Bay, FL 32905 USA

4164 Lakeshore Road, Burlington,
ON, L7L 1A4 Canada

CRC Press
6000 Broken Sound Parkway NW,
Suite 300, Boca Raton, FL 33487-2742 USA

2 Park Square, Milton Park,
Abingdon, Oxon, OX14 4RN UK

Library and Archives Canada Cataloguing in Publication

Title: The food and beverage hospitality industry in India : an emergent segment / Sandeep Munjal, Sanjay Sharma.

Names: Munjal, Sandeep, author. | Sharma, Sanjay (Chef), author.

Description: First edition. | Includes bibliographical references and index.

Identifiers: Canadiana (print) 20210162899 | Canadiana (ebook) 20210163127 | ISBN 9781771889094 (hardcover) | ISBN 9781774638040 (softcover) | ISBN 9781003034544 (ebook)

Subjects: LCSH: Hospitality industry—India. | LCSH: Food service—India. | LCSH: Beverage industry—India. | LCSH: Food consumption—India. | LCSH: Food habits—India.

Classification: LCC TX910.I5 M86 2022 | DDC 647.940954—dc23

Library of Congress Cataloging-in-Publication Data

Names: Munjal, Sandeep, author. | Sharma, Sanjay (Chef), author.

Title: The food and beverage hospitality industry in India : an emergent segment / Sandeep Munjal, Sanjay Sharma.

Description: First edition. | Palm Bay, FL : Apple Academic Press ; Boca Raton, FL : CRC Press, 2021. | Includes bibliographical references and index. | Summary: "This book, The Food and Beverage Hospitality Industry in India: An Emergent Segment, presents the story of growth and change of what is still a largely unorganized food and beverage service industry in India. With the authors' experience in both industry and academia, the volume provides a holistic perspective of the current performance and identifies the topical issues and the challenges faced by the Indian food and beverage industry. The authors offer an insightful discussion on where the industry is headed and how it can move from top-line driven growth to a bottom-line supported one. The authors cover changes over the last two decades in food and beverages services in India, profiles of Indian consumers, also looking at sociocultural as well as economic diversity. Issues and challenges faced by vendors and customers, such as the level of and need for skilled and trained manpower, lack of policy support, over-regulation, etc. The tremendous opportunities to grow and the need for businesses to achieve profitability. Business models that drive investments, the specifics of service-related front end of the business, as well as the culinary back end. How technology is changing the way business is being done and the rich culinary heritage of India and current food trends. This volume will be a valuable resource for hospitality industry managers, students of hospitality programs, hospitality sector entrepreneurs and owners of hospitality businesses, as well as professionals working in the areas of sustainability, slow food, farming businesses, and others"-- Provided by publisher.

Identifiers: LCCN 2021011441 (print) | LCCN 2021011442 (ebook) | ISBN 9781771889094 (hardcover) | ISBN 9781774638040 (paperback) | ISBN 9781003034544 (ebook)

Subjects: LCSH: Food industry and trade--India. | Beverage industry--India. | Hospitality industry--India.

Classification: LCC HD9016.I42 M828 2021 (print) | LCC HD9016.I42 (ebook) | DDC 647.9554068--dc23

LC record available at https://lccn.loc.gov/2021011441

LC ebook record available at https://lccn.loc.gov/2021011442

ISBN: 978-1-77188-909-4 (hbk)
ISBN: 978-1-77463-804-0 (pbk)
ISBN: 978-1-00303-454-4 (ebk)

About the Authors

Sandeep Munjal

Sandeep Munjal is a Director at Vedatya Institute, Gurgaon, in the National Capital Region (NCR Delhi), India. Professor Munjal brings a unique combination of extensive industry and academic experience spanning over 23 years. Having worked with industry leaders like the Taj Group of Hotels in India and ARAMARK Corp. in USA, he offers a global perspective. He has also been engaged in a range of consultancy assignments in hospitality and retail space in India. An IHM Aurangabad, alumni, awarded the best all round performance medal, Prof. Munjal also pursued higher education in USA in hospitality management at Widener University, Chester, PA. He is an avid researcher and has published widely in the area of hospitality management and heritage tourism in international and national journals of repute. Throughout his career journey, he has received numerous awards and certifications.

Sanjay Sharma

Chef Sanjay Sharma heads the school of culinary excellence at Vedatya Institute, Gurugram, in the National Capital Region (NCR Delhi), India. He is a certified hospitality professional from Oxford Brookes University, UK, and a certified food handler from Whitbread Academy, Birmingham, UK, and he has more than 17 years of work experience in the hospitality industry and academia. A graduate from the Institute of Hotel Management, Lucknow, he brings rich national and international work experience. He upgraded his education and honed his management skills through an MBA program at Birmingham City University, Birmingham, UK, and subsequently worked with Whitbread, UK, in various positions. Some of the prominent names that he was associated with include Taj Hotels, The Park Hotels, IHM PUSA, IIMT Oxford Brookes University, Premier Inn, UK, and Whitbread, UK. Chef Sharma is an avid researcher and has extensively published nationally and internationally in the areas of culinary heritage, sustainability, gastronomic tourism, slow food, and food cost control.

Contents

Abbreviations

ACDR	affordable casual dining restaurant
BYOB	bring your own booze
CAC	Codex Alimentarius Commission
CDR	casual dining restaurant
DOFM	Delhi Organic Farmers Market
EIh	East India Hotels
EOO	eating out occasion
F&B	food and beverage
FD	frozen desserts
FDI	foreign direct investment
FDR	fine dining restaurant
FSMS	Food Safety and Management System
FSSAI	Food Safety and Standards Authority of India
GDP	gross domestic product
GST	goods and services tax
HACCP	Hazard Analysis and Critical Control Points
IBEF	India Brand Equity Foundation
IC	ice cream
IHCL	Indian Hotels Company
KOTs	kitchen order tickets
MRL	maximum residue level
MT	metric tons
NASVI	National Association of Street Vendors of India
NOC	no objection certificate
NRAI	National Restaurant Association of India
NSDC	National Skills Development Corporation
PBCL	pubs, beer café's and lounges
PCDR	premium casual dining restaurant
POS	point of sale
QSR	quick service restaurant
SME	small scale enterprise
SOPs	standard operating procedures
VAT	value-added tax

Preface

The journey of the Indian economy in the post-liberalization era of the 1990s has been an area of interest for a wide range of stakeholders globally. This includes investors, businesses across multiple sectors, governments, and customers. Over the last two decades, it has been impossible to ignore the Indian growth story and its immense capacity to become the quintessential engine of global economic growth, a beacon of hope, so to say, in a world where the industrialized economies have capped off their economic growth such that they are either growing at weak GDP growth rates or even worse dealing with periods of economic downturn. The size of the Indian market in terms of the number of consumers, the improving purchasing power of this massive customer base with ever-rising disposable income, and a desire to spend is music to ears of businesses anywhere in the world. The demographic dividend of a young population (forecasted median age of 29 by 2030) in a world where most nations are grappling with the realities of aging populations makes India even more interesting as well as critical to the strategic choice of businesses globally. The current Indian government has announced an agenda to ensure that India becomes a $5 USD trillion economy by 2025. The government has a massive investment-oriented blueprint in place to make this happen. Service industry at large is expected to play a major role as the Indian economy gradually continues to shift from an agriculture-centric orientation to one with a strong contribution from the service and manufacturing sectors. This brings tourism and hospitality industry into play like never before.

According to the National Restaurant Association of India (NRAI), food and beverage service in India is expected to be nearly INR 5000 billion in revenue terms, expected to grow at a rate of anything between 8 and 10% through the 2016–2025 time period. It is relevant to talk about these statistics as one gets an overview of where India is headed in terms of the broad economic sense as well as how the food service industry is placed within the larger ecosystem. Globally the impact of the tourism and hospitality sector as a driver of economic growth, with a huge potential to generate direct and indirect employment, is well recognized. This opportunity also attracts political attention, and India is no different in that respect.

Governments at both the Center as well as various state governments in the country have made the push through policy action as well as legislation to ensure that industries associated with these sectors thrive and receive support in the form of public–private investment such that the political dividends in the form of employment generation can be unlocked.

While there has been ample published research with respect to the tourism sector and hotel-focused aspects of the hospitality sector in India, the massive food and beverage service industry has remained ignored as far as Indian experience of its growth and transformation is concerned. Led by NRAI there has been an effort to collect data on key metrics around the food service industry in India and to provide a voice to the issues and challenges of the industry. While industry organizations like the FHRAI (Federation of Hotels and Restaurants Association of India) and FICCI (Federation of Indian Chambers of Commerce and Industry) have from time to time covered the F&B service industry in their annual reports or special features, the coverage has been far from comprehensive, and with respect to FHRAI, it has been clearly largely about hotel businesses in India.

It is this gap that got us (authors) interested in writing this book, which is completely focused on the food and beverage service industry in India. The lack of research focus on the industry was also evidenced in the form of limited secondary sources of data beyond what was made available by NRAI. As the mouthpiece of the Indian food and beverage industry representing more than 500,000 restaurant businesses with revenues over USD 4 billion, the association has offered credible statistics that this book has referred to widely.

The book is focused on presenting the story of growth and change of what is still a largely unorganized food and beverage service industry in India. The authors have made an effort to provide a perspective on both the service-related front end of the business as well as the culinary backend so that the readers have a holistic perspective of the industry rather than a one-sided view. The journey of the F&B service industry over the last two decades has been quite fascinating, and in many ways a result of the economic liberalization that commenced in 1993 under the leadership of Dr. Manmohan Singh who subsequently also headed two UPA (United Progressive Alliance) governments at the Center. The earlier mentioned interest in the tourism sector quite easily translated into a focus on the hotel industry where the market share was largely divided between three domestic players, namely IHCL (Indian Hotels Company Limited) or the Taj Group, EIH (East India Hotels) or the Oberoi Group, and ITC Hotels (owned by the Indian Tobacco Company). This was soon to change as opening doors of FDI (foreign direct investment) saw every major international hotel brand find presence in India, you name it and the brand was there competing for market share or rather expanding the market itself. This transformation of the hotel industry was taking place under the backdrop of many other socio-cultural changes in India, if not more, equally relevant to the economic changes mentioned earlier. A diverse, large country was changing, changing in terms of travel habits, consumption of food, and slowly but surely building an appetite to experience hospitality. This shift, of course, had everything to do with an improved economic situation, higher employment, growing wages, more double-income families with more disposable income than ever before in the young nation's history.

As hotel businesses grew, the growth came in terms of number of keys across cities and number of segments (luxury, mid segment, and budget segment) with both Indian and international companies committing investments over the two decades. A parallel story was emerging alongside, one that saw the Indian metropolitan cities welcome the arrival of air-conditioned cafes, international fast food brands opening restaurants, and more full service restaurants being opened much to the glee of the Indian consumers of all ages. The book charts all facets of this growth story. It is not lost to the authors that the food and beverage service industry still is a lot about the huge "unorganized formats" represented by the roadside "dhaba," "nukkad tea stall, " street food on "thela, " local "mithai" shop cum eatery, and thousands of simple, non air-conditioned eating places that can be found all over the country. This segment still constitutes nearly 60% of the overall food and beverage service market. The arrival of organized retail formats in India has made an impact but the real threat to unorganized local "kirana" stores or grocery shops did not come from the "brick and mortar" store formats, but rather from the online discount retailers. As the organized food and beverage service formats continue to grow, it would be interesting to see how the unorganized formats respond. Will it be a "melting pot" in which the traditional setups absorb the differentiators of the organized players and evolve into them retaining their identity and profiting from the professional business approach of the organized formats? Too early to call, but it looks like the consumers will call the shots. As Indian consumers demand better service, better hygiene standards, better ambiance, and more cuisine variety, the businesses will have no choice. They will either respond to these expectations and change or become unviable. This is not to suggest that traditional, ethnic, local food will not be in demand, in fact to the contrary there seems to a renewed interest in all that is traditional and connected to our roots. That said the urban–rural divide is still wide and the change in the rural and semi-urban Indian hinterland is going to be much slower than the rapid shift that was witnessed in the metropolitan and tier 1 Indian cities. The tier 2 and 3 cities and smaller towns are going to witness the next push of expanding F&B service retailers. All of these would also continue to be affected by global trends and the impact of technology at large.

The authors have made an effort to provide readers a strong perspective on the above-mentioned aspects in a focused manner through an array of ten chapters, each looking at a specific aspect of the food and beverage service business in the Indian context. Chapter 1 clearly sets the tone by discussing how the hospitality industry has changed over the last two decades from one where hotels worked with a "room revenue" mindset to one today where F&B segment is where significant efforts to innovate and bring in differentiation is emerging. The chapter also establishes that the growth of organized food and beverage service formats in India has reached a stage where the hotel based F&B service business in size is reduced to a less than 5% of the overall organized F&B service market. This is in sharp contrast to the pre liberalization era India, where the only organized F&B outlets that existed were ones based at the luxury hotels. Chapter 2 takes a look at the organized F&B service market in terms of the subsegments that have found roots over the last two decades. The classification offered by NRAI is useful here as the various service formats like cafes, bar cafes, casual dining restaurants, premium casual dining restaurants, quick-service restaurants, fine dining restaurants, and ice cream bars are discussed in terms of their growth as well as respective market shares. It becomes quite apparent that it is largely the quick service and casual dining formats that are controlling nearly 70% of the organized F&B service market, another aspect that the chapter highlights is the response to branded chain based businesses as against the nonchain standalone. The standalone businesses seem to be doing well and have nearly two-thirds of the market share.

Chapter 3 is focused on the typology of the Indian customer and also tries to understand their buying behavior in terms of reasons to buy or eat out and also look at with whom they eat out. The demanding nature of the Indian consumer is quite apparent as the need for value is significant. The NRAI typology is used as a framework by the authors as they identify four types of consumers—family bonding seeker, fun seeker, socialize, and the discerning Indian. The segments can be identified by varied typologies, but this one drives home the socio-cultural as well as economic diversity that India represents. The chapter attempts to answer all key questions that can explain the typical Indian's buying behavior with respect to F&B service products. Chapter 4 takes a macro view of the F&B service industry in India and highlights the major issues and challenges. Be it the issue of skilled and trained manpower or lack of policy support, judicial overreach, and aspects like overregulation, the author discusses these aspects in detail and also offers recommendations on how these issues and challenges can be resolved. The tremendous opportunities to grow are identified and also the need for businesses to achieve elusive profitability is amply discussed.

Any discussion on a high growth industry is incomplete without due review of the "business models" that are driving investments and finding acceptance, Chapter 5 does just that. The author looks at both the front end as well as back end (infrastructure, logistics, supply chain, and cold chain) related investments that have driven the growth and expansion. It is quite apparent that the "franchise" model is where maximum play has come from. Private equity, funds based (domestic and international) as well as venture capital-based funding has been quite successful. This holds true for both the front end as well as the back end of the business. Many corporate too have made forays into the F&B service business as a diversification strategy of sorts. As mentioned earlier the top-line growth has so far not been complemented by the bottom line, this seems to be a problem that the industry must find an answer to as profitless growth is hard to sustain. The role of technology in changing the way business is being done is hard to ignore, such is the sway of technology that it has potential to change the operating models for food and beverage businesses in India.

Chapter 6 is focused on the very interesting facet of the F&B service industry in India that recognizes the rich culinary heritage that country possesses and how it is critical for the organized formats to not lose sight of the value proposition. While traditional food, slow food, and food of the land (local food) are popular aspects for the restaurant industry globally, the Indian industry is now moving in that direction. Restaurant operators are making efforts to offer menus that give space to traditional and local food. The author also shares examples of initiatives on the ground focused on strengthening traditional/slow food; including preventing loss of traditional recipes and by default our culinary heritage. The next chapter picks up from here and provides ample inputs on the type of trends that the industry is witnessing currently, especially the focus on fresh ingredients locally procured to prepare authentic food that tastes like "gharkakhana," food like that cooked at home. The chapter does highlight some trends around the food to look out for.

Chapter 8 looks at efforts on the ground by the food and beverage service industry to make efforts toward backend integration. While these efforts may have found strategic value initially due to high food inflation, over the years demand of fresh organically produce as well a hard push toward finding sustainable solutions seems to be the driver of efforts now. The author again shares examples of practices by businesses as well as educational institutions to further the cause. Chapter 9 is about the transformation of Indian street food in India. It discusses the issues and challenges faced by the vendors, further the discussion is on popular Indian snacks and ways to improve the infrastructure and challenges faced by the vendors and customers. Chapter 10 loops in the response of the F&B service industry in terms of "kitchen design" elements as well as a brief coverage of how the organized formats have made efforts to offer value to the customers by redefining the aesthetics and other aspects effective service side. The unique response of chain and nonchain formats in this regard is interesting. The book signs off with a tribute of sorts to the street food of India that has defined the nation's cultural, religious, and regional diversity. The author also offers a view on how the commercial expression of the country's culinary heritage is where a huge value proposition exists. It will be interesting to see if the organized formats in F&B space take it upon themselves to unlock this massive potential.

We sincerely hope that the readers are able to gain useful insights into the food and beverage industry in the Indian context. It has been our endeavor to offer a holistic view into the current status of the operators in the key segments or formats, and also shares some pointers in terms of where the businesses are headed given the unique socio-cultural-economic setting that India represents.

—Sandeep Munjal
Sanjay Sharma

CHAPTER 1

Indian Food and Beverage Services Segment at a Glance

ABSTRACT

The chapter opens the conversation about the dynamic food and beverage service space in India, setting the tone of the book as it attempts to present the current status of the segment and way forward. On one hand, the hotel industry itself in India has warmed up to the idea of pushing revenue growth through focus on F&B, investments (domestic and international) are fueling the growth of organized food and beverage service retail formats outside of the hotel industry on the other. Socio-cultural-economic factors are at play as a wide range of food and beverage formats find consumer interest. The chapter underpins the reality that while the unorganized segment of the food and beverage service industry remains large, the momentum is clearly toward professionally managed chain or nonchain organized operations. The sheer number of start-ups in the food and beverage service industry in India, over the last decade, makes quite a statement.

1.1 INTRODUCTION

The decade's postliberalization of the Indian economy in the early 1990s saw the hospitality sector push forward with investments both domestic and FDI (foreign direct investment) driven, as a result over the next few years nearly all global hotel brands had established presence or strengthened their existing presence in the fast growing Indian market. Though a comprehensive national tourism policy was announced in 1982, it was only by 1988 that the government was able to come up with a comprehensive tourism promotion plan. It was hereafter that the industry received the policy focus that it had been waiting for. From 1990 onwards, the decade saw most Indian states come with policies that supported tourism growth at the respective state level and as a direct result started to look at the hospitality industry as a driver of tourism-related economic growth. Public–private partnership became the new mantra and in 2002 when the revised tourism policy was announced which brought the need for infrastructure development in focus, this included accommodation, roads, airlines, airports, food services, and more.

The impact of the above mentioned favorable policy action is now visible at multiple levels, from a macroeconomic perspective, in 2017 hospitality and tourism accounted for 9.4% of the GDP (gross domestic product) and was the third largest contributor to foreign exchange earnings (IBEF.org). The India Brand Equity Foundation (IBEF) report 2017 also states that growing at a compounded growth rate of 14% for the last decade, the travel and tourism sector is expected to directly contribute USD 150 billion or more to the Indian GDP by 2027. It is also being forecasted that the sector's overall contribution to GDP will reach 10% at nearly USD 425 billion by 2027.

The above-discussed aspects of the travel and tourism growth story in India is amply supported by investment, this includes domestic (including that made by the central and state governments) and also substantial amounts of FDI inflows largely in the form of hotel/hospitality related projects with international fund or venture capital funding. The policy decision to allow up to 100% FDI investment through the automatic route, adhering to established procedures and other regulatory requirements, has made a key difference. Hotel and tourism saw cumulative FDI inflows of nearly USD 11 billion from 2006 to 2017, with hotels, resorts, and recreational infrastructure being the prominent areas to attract these investments. As a welcome change, nearly half the ministry of tourism planned budgetary outlay found utilization toward infrastructure addition and upgrade spread over the last decade or so.

JLL (Jones Lang LaSalle) in their publication, India Hospitality Industry Review (2017), provides a well-articulated narrative on the recovery of the Indian hospitality sector from the period of slow GDP growth in 2015–2017 due to fiscal decisions like demonetization and implementation of the goods and services tax (GST) by the central government. According to the report, as the uptick in the manufacturing sector and the construction sector (post real estate regulation act) kicks in the GDP growth rate is likely to move north further, hitting a close to 8% bar in the 2018–2020 period. The hospitality sector seems to have moved past some not so supportive policy actions that were taken by the government including the liquor sale restrictions on hospitality businesses located within 500 m of any national highway, as also the issue of high GST (goods and services tax) slabs as well as the applicability of the taxes on published hotel tariffs.

The ministry of tourism has reported a strong 16% growth in foreign tourist arrivals (FTA) and the absolute number has crossed the 10 million mark for the first time. hotels have reported either better ADRs (average daily rate) or occupancy or both, indicating that the recovery is strong and is likely to maintain speed through 2018–2019. Figure 1.1 articulates the trend discussed.

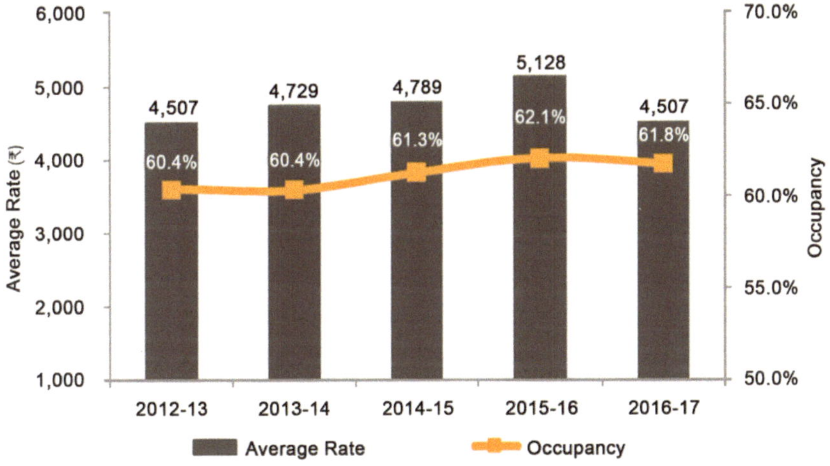

FIGURE 1.1 Hotel occupancy and average rate (2012–2013 to 2016–2017).
Source: Indian Hotel Industry, FHRAI Report 2017.

A slower growth in new hotel openings has also helped the cause as the demand–supply gap has been closed significantly. The JLL report (2017) offers an insight into the supply side of things by citing data that establish that in 2016 and 2017 the number of brand signing remained in the 170 odd hotels range, amounting to an addition of 16,000 keys. The headline capturing trend of 2017 was the number of hotel conversions that were recorded. In all, 62 conversions took place in 2017 as against 33 in

2016, largely as a result of Accor losing out to IHG (Intercontinental Hotel Group). Another noteworthy trend has been the shifting of gears with respect to the fresh hotel signings by the domestic (Indian) hotel players as against the international brands. The year 2017 saw the domestic brands sign 56% of the signed up capacity, reversing a previous trend sharply. This has implications for the sector, as the domestic hotel brands are showing a strong appetite to join the capacity expansion party. Domestic or International brand driven, the spread of the upcoming capacity addition seems to be quite evenly spread across the tier 1, 2, and 3 Indian cities.

From a perspective of the type of segments finding investment interest, the Indian hotel industry seems to be finding affinity toward the leisure driven segment, though commercial/business hotel signings at about 65% are still the biggest driver, at 30% the leisure segment is showing renewed promise. The commercial hotel demand is also quite closely linked with the "office space" market, JLL report (2017) establishes that nearly 28 million square feet office capacity was absorbed in 2016 and an additional 27 million in 2017, as the supply side eases off the office market will stabilize even further, a positive indicator for the Indian hotel industry at large and specifically the commercial hotels.

There is however other evidence that all is still not well with the macroeconomic elements; early 2019 has some worrisome indicators to watch out for; the Indian stock markets have been witnessing weakness emanating from poor corporate results across multiple sectors; and multiple other indicators of economic health like the unemployment rate to look rather unhealthy. That said a strong second mandate in the 2019 general elections for the BJP (Bhartiya Janta Party) government at the centee does bring continued political stability to handle whatever is headed (livemint.com, 2019). Despite the challenges, it is quite apparent that the Indian hotel industry is chugging ahead, having shrugged off the impact of any macroeconomic deterrents that may have slowed down its growth over the last few years. The positive impact of the tourism sector related expansion in terms of services, facilities, and infrastructure elements mentioned earlier is being clearly witnessed. The India Hospitality Review (2018) report from HVS offers predictions for 2019–2020; according to the report, Hotel RevPAR (revenue per available room) should grow by 9.5%, largely on the back of growth in Average Daily Rates (ADR). The gross operating margins are likely to improve by 8.0% and as mentioned earlier supply is anticipated to grow slowly at 4.0% or even lower while demand growth should hold steady at 7.0%. But to put more perspective on these encouraging numbers, the report also mentions that in rupee terms even at this rate of growth the Indian hotel industry will take another 5 years to come close to the returns that were witnessed in the boom period before the bust in 2009.

Some of the key new trends with respect to travel and tourism are shared as given below.

Emerging tourism trends are:

- Wellness tourism
- Adventure tourism and camping
- Cruises
- Spiritual tourism

1.2 IS INDIAN HOSPITALITY BUSINESS STILL ALL ABOUT ROOM REVENUE?

A fair question to ask, if we were to trace the origins of the Indian Hospitality sector we would easily mark December 16, 1903 as a pivotal moment with the Taj Mahal Hotel, Mumbai opening its doors to the guests in pre-independence India. As the country gained independence, the hospitality sector saw the emergence of additional competitors in the form of East India Hotels (EIH) and later with ITC (Indian Tobacco Company) making investments to build and operate hotels. The economic liberalization of the 1990s as discussed earlier saw foreign investment in the form of international brands arrives in India. Soon enough most of the prominent brands including Hyatt, Marriott, Carlson Rezidor (now Radisson Hotels Group), Hilton, and IHG (Intercontinental Hotel Group) were operating hotels in India. Initially, these investments were in the luxury segment alone and focused on tier 1 cities (the metro cities). Eventually, the growth expanded to the tier 2 and tier 3 cities as well. Another facet of this growth has been the interest in the mid-segment as well as the budget categories. This allowed international brands to also expand their brand portfolio in India further.

The domestic brands did the same after recognizing the potential of the segments. In the absence of any pre-existing big room inventory in the 4-star/3-star hotel segment, the industry saw a rush in terms of new hotel signings. Many new domestic mid-segment hotel brands emerged, Lemon

Tree Hotels, Golden Tulip, and Sarovar to name a few. Over the last two decades, this has resulted in a build-up of significant room capacity across all star categories, including the heritage classified hotels as mentioned in the Table 1.1 below. While the average size of the room inventory that a given star category currently has continued to be significantly lower than what would be the case in the developed economies, the overall room capacity, as well as the room inventory per hotel, has been growing consistently.

1.2.1 INDIAN HOTEL INDUSTRY—F&B SERVICE FACILITIES ANALYSIS

Table 1.2 provides a parallel input on the average number of food and beverage outlets per Hotel across the cited categories. It is quite apparent that larger the room inventory more the number of food and beverage service outlets at a hotel. While this statistic is representative of the space and investment toward food and beverage service points of sale, what is more important to get a perspective on is the revenue spread that the hotels are witnessing today between room revenue and F&B revenue. Historically, hotels in India were all about room revenue and revenue from other sources including food and beverage played second fiddle. This aspect of the hotel industry saw a shift over the last decade or so as ARRs started to flatten off and food services emerged as a potential differentiator and reason for choosing one hotel over the other.

The dwindling focus on rooms has given way to a new found interest in food and beverage driven revenue growth in recent years. The fact that the "room revenue" came at a much lower cost as against any other revenue source at an operating margin level cannot be debated. As a result potentially till the end of the century, the dominance of the "rooms" segment remained unchallenged. Not surprising as nearly 65% of the revenue of a typical hotel was from rooms, and the remaining from F&B and other sources.

Recent years have seen that changes in F&B revenues have grown (see Fig. 1.2 below) to over 40% of the total revenue.

TABLE 1.1 Typical Room Profile of an Average Hotel in India.

Star categories										**Room count**		
Composition	**Five-star deluxe**	**Five-star**	**Four-star**	**Three-star**	**Two-star**	**One-star**	**Heritage**	**Others**	**2016–2017 all India average**	**<50**	**50–150**	**>150**
Air-conditioned												
Rooms	219.5	135.6	99.0	52.9	34.0	21.0	21.5	30.6	62.4	25.5	74.9	221.2
Suites	11.4	9.1	8.9	4.4	2.9	1.3	9.2	2.2	5.0	2.2	6.6	15.0
Non-air-conditioned												
Rooms	0.5	1.4	0.0	2.1	3.5	6.7	2.7	3.4	2.4	2.9	2.2	0.2
Suites	0.0	0.1	0.0	0.2	0.2	0.6	1.1	0.6	0.3	0.3	0.2	0.0
Total average rooms	231.4	146.2	107.8	59.5	40.7	29.5	34.5	36.9	70.1	31.0	84.0	236.4

Note: The number of non-air-conditioned rooms in lower star category hotels appears to be higher, partly because many are located in hill stations where air-conditioning is not provided in hotels.

TABLE 1.2 Average Number of Food and Beverage Outlets per Hotel in India.

Star categories										**Room count**		
Composition	**Five-star deluxe**	**Five-star**	**Four-star**	**Three-star**	**Two-star**	**One-star**	**Heritage**	**Others**	**2016–2017 all India average**	**<50**	**50–150**	**>150**
Restaurants	3.4	2.5	2.0	1.6	1.3	1.3	1.4	1.4	1.7	2.4	1.8	2.9
Bars	1.5	1.1	1.0	0.7	0.5	0.4	0.7	0.5	0.7	0.6	0.8	1.3
Others	0.6	0.7	0.5	0.3	0.2	0.2	0.4	0.1	0.3	0.2	0.4	0.6
Total	5.4	4.3	3.5	2.6	2.0	1.8	2.4	2.0	2.8	3.1	3.0	4.8

Source: Indian Hotel Industry, FHRAI Report 2017.

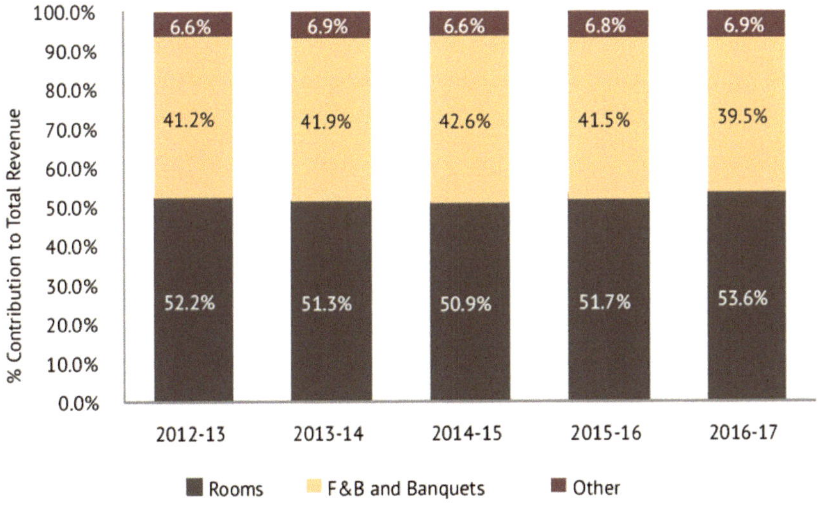

FIGURE 1.2 Shift in revenue spread between "rooms" and "F&B".

Source: Indian Hotel Industry Survey 2016–2017, FHRAI (2017).

The reasons for this shift in focus is a little more complex given the fact that capacity addition across the hotel segments over the last few years saw an oversupply situation emerge. As a result, ARRs (average room rates) either declined in some market segments or flattened off, hence, a scramble by all players to look for opportunities to build revenue growth by expanding the footprint of food and beverage services.

1.2.2 ROOMS VS F&B OUTLETS—WHERE IS THE DIFFERENTIATOR?

Recognizing that economic slowdowns too are cyclic and thus pass one may also argue that the supply side discussion aside, the room product for each star category across various competing brands in the Indian market has started to become similar with less and less to segregate one from another. The cookie cutter approach to designing the hotel rooms in terms of their size, look, and feel seems to be the causative factor. The market leaders have invested in technology to bring the new age "room" to the new age Indian, only to find that competition (domestic chains and international brands) alike following suit resulting in any room product-based differentiation a short-lived competitive advantage. Attracting price premium based on the room product has indeed become challenging in the digital era where it is the 'practice' of a revenue management strategy that currently guides the pricing strategy more than anything else. Room inventory is being sold mostly online from a range of channels/platforms to a customer base that is more price loyal than brand loyal. This seems to be the hard reality on the ground. As mentioned by Singh and Munjal (2012), the sway of technology is massive, while it has made room rate transparency the new normal. It has also opened doors for room aggregators like OYO and others to find roots. This environment has turned the "selling function" at Hotel brands on its head. The sales teams across all brands have had to formulate a strategy that is not "room centric" anymore, but now looking at the **"food and beverage" product as a potential differentiator**, it is this strategic shift in thinking that has led operators to invest more enthusiastically in building food and beverage related infrastructure including banquet and meeting spaces.

It is clear that food and beverage product and service is today central to any strategic choice being exercised by a hotel brand in India to achieve sustainable competitive advantage.

F&B focus is the new strategic imperative for Hotel Industry in India.

1.3 EVOLUTION OF THE NON-HOTEL-BASED F&B SPACE IN THE INDIAN MARKET

The journey of Indian hospitality sector can be traced to pre-independence as mentioned earlier in this chapter. The food and beverage component of hospitality charted a growth pattern alongside the growth of the "hotel" based hospitality sector, in parallel the unorganized, unbranded restaurants, sweet shops, and street food options existed and supported the demand for 'F&B' products and services in every nook and corner of the country, often expressing the socio-cultural diversity of the region/city/town/locality in the form of the products made available for sale and how they were served. Sharma (2012) in his research on traditional Indian food establishes the presence of huge, marketable culinary heritage akin to the built heritage that is a quintessential driver of tourism activity (both domestic and international). Given the immense diversity in our culinary resource, it is no surprise that the hotel industry has turned to the food and beverage segment to achieve competitive differentiation, revenue growth, and enhanced profitability.

The NRAI report (2016) documents the "food services" market in 2016 in India to be worth USD 48 billion in revenue terms, while Hotels as an industry is mentioned to be USD 6.3 billion as shared in the table below. The shared data (Table 1.3) offer not only a perspective on the market size of this industry but also provide a comparative perspective of sorts to reflect upon. For a country with a population size that is 1.2 billion and a mammoth middle class with expanding spending power, the market size growth is likely to continue at a feverish pace. With improvement in travel-related infrastructure that is the guiding growth in domestic tourism numbers, the food and beverage service/retail expansion is only logical.

TABLE 1.3 Industry-Wise Market Size in India.

Industry	Market size 2016 (USD billion)*
Retail	688
Insurance	60
Food services	48
Pharmaceuticals	23
TV	12
Hotels	6.3
Films	2.3

* As per prevailing INR to USD currency conversion rates.

Source: NRAI Report, 2016.

1.3.1 F&B SERVICE MARKET: SEGMENTS AND AFFILIATION

The earlier discussion on the overall market size of the food and beverage service businesses in India would be incomplete without offering perspective on the prominent segments and the overall structure of the market in terms of standalone businesses as against larger "chain affiliation" based formats that would include franchising as well as owned business models. It is critical to understand that food and beverage service operations in the organized format are no longer limited to the confines of a hotel. In fact, a large part of the growth story in F&B services has been outside of the traditional hotel setting, bringing organized food and service retail at the center of the discussion.

The retail space in general in India has remained overwhelmingly "unorganized" with nearly 88–90% of business in the hands of the small retailer. While organized formats in the "brick and mortar" store format have made a dent over the last decade, the same barely compares to what is typical of any developed economy where 90% or more of the retail market is with the branded, organized players. The online retail space however has been expanding rapidly with domestic startup like Flipkart finding a buyer more recently in Walmart. The market share fight is now largely between the global biggies Amazon and Walmart. The reason why it is important to understand the retail space in general is because organized "food and beverage" retail segments are likely to take a cue from the larger and broader trends that the sector is witnessing in India.

The large part of the Indian food and beverage outlets/operations are closer to what Figure 1.3 showcases: small, non-airconditioned, serving Indian food, managed by a team not offering the look and feel of trained professionals yet packing a punch with the food, and HACCP aspects notwithstanding. That however is changing and changing fast. Prakash (2004) in his widely acclaimed work on cultural history of India offers a detailed perspective on how India and her "food" are inseparable. While the richness of the "culinary" knowledge cannot be underplayed or ignored, it is the front end of the service that needs to come in terms with the modern day customers. Figure 1.4 offers a visual perspective of how organized F&B retail has found interest of consumers in urban India.

FIGURE 1.3 A typical unorganized format F&B outlet in India.

Source: Author.

FIGURE 1.4 Changing face of F&B service retail in India.

Source: Author.

The food and beverage space is complex in terms of how it can be categorized. F&B services exist in the form of outlets managed within the larger "hotel" operations context as well as independent units without the accommodations aspect in tow. The segmentation of the F&B service/retail space can be done in varied ways. The data shared by NRAI (2016) in their report do a good job in terms of providing inputs that are categorized into formats and then further dissect the market share into chain-based operations and stand-alone businesses in an organized space or otherwise. The segment below also offers perspective on the average spends that can be expected from a typical customer for each of the mentioned segments in Table 1.4.

TABLE 1.4 Structure of Indian Food Services Market.

Key segments in the food services market	Average spend per person (INR)
Unorganized segment: It includes roadside eateries and "Dhabas" which have been the most common eating out option for decades in India. these too have evolved as per customer expectations, largely confined to the state and national highways and the rural parts of the country.	10–100

TABLE 1.4 *(Continued)*

Key segments in the food services market		Average spend per person (INR)
Organized segment: Consists of: (a) Standalone restaurants across all formats with less than three outlets. (b) Chain format which has three or more outlets across all formats.		
Chain segment		
Café	Coffee and chai bars as well as parlors and bakeries. Strong focus on beverages supported by food item, e.g. • Starbucks, Café Coffee Day, Costa Coffee, Indian Coffee House, Barista, etc.	50–250
Quick service restaurants (QSRs)	Focused on speed of service, affordability, and convenience. Strong focus on take away and delivery with minimal table service, e.g., • Haldiram's, McDonald's, KFC, Harish, Bikaneriwala, Burger King, Subway, Dunkin Donuts, Burger Singh, etc.	75–250
Frozen desserts/Ice cream (FD/IC)	Comprises small kiosk formats of ice cream brands and has now extended the dine-in concept to frozen yogurt brands, e.g., • Baskin-Robbins, Keventers, Red Mango, etc.	50–150
Affordable casual dining restaurants (ACDRs)	A restaurant serving moderately priced food in an ambience oriented toward providing an affordable dining experience with table service. The offerings bridge the gap between QSRs and premium casual dining restaurants, e.g., • Pind Balluchi, Sagar Ratna, BBQ Nation, TGI Friday, Ruby Tuesday, etc.	250–500

TABLE 1.4 *(Continued)*

Key segments in the food services market		Average spend per person (INR)
Premium casual dining restaurants (PCDRs)	Restaurants bridging the gap between ACDRs and fine dining restaurants. These are full-service restaurants with eclectic high quality interiors and high standards of service, e.g., • Farzi Cafe, Oh! Calcutta, etc.	500–1000
Fine dining restaurants (FDRs)	A full-service restaurant with premium interiors, specific cuisine(s) specialty, and high standard of service. They offer a unique ambience and an upscale service with the help of highly trained professional staff, e.g., • Copper Chimney, Olive Bar, etc.	>1000
Pubs, bar café & lounges (PBCL)	This format mainly serves alcohol and related beverages and includes night clubs and sports bars, e.g., • Down Town, Beer Cafe, Xtreme Sports Bar, etc.	750–1500

Source: NRAI Report, 2016.

The above-mentioned table provides perspective on the spread of the food and beverage service space in the Indian market currently. The unorganized and unbranded segment remains dominant and quite massive; however, the available statistics on the number of such outlets and the market share need to be taken with a pinch of salt as varying sources would offer different data. What is quite apparent is that over the last two decades the food and beverage service market has not only just grown in terms of its revenue impact but also acquired a width of market segmentation that is indicative of a maturing industry. From the roadside eateries, dhabas, mithai shops, and street food joints that represent the unorganized segment to the organized segment that includes both chain and non-chain variants, the customer's choice and purchasing power have encouraged the food and beverage service market to segment itself such that there is an offering available to fit every budget based on the product-service duo variability.

The impact of an expanding economy, globalization, and a shift in socio-cultural factors have all had an impact on the consumer expectations and demand for food and beverage products and services. A key driver of the demand for varying subsegments like the QSRs (quick-service restaurants), casual dining, fine dining, premium casual dining, cafes and pubs, etc., is the burgeoning Indian middle class. With stronger purchasing power, disposable income, and a will to travel, Indians are spending more and more on dining experiences. Dining out of the house or other alternatives like drive in/take away/home delivery have found a strong demand through the last two decades. The below cited data from the NRAI, 2016 report (Table 1.5) provide a broader perspective on how the consumption expenditure panned out for both the nondiscretionary as well as discretionary elements. The nondiscretionary spend on food and grocery increased by 36% from 2013 to 2016, the same is expected to grow more than 60% by 2021. This, however, is not of particular interest from the context of this book; the focus really is on the discretionary spending growth specific to the food services segment. A nearly 20% growth in consumption expenditure from 2013 to 2016, and a forecasted nearly 61% growth from 2016 to 2021 sets the stage for a robust demand side equation for the Indian food and beverage service industry to consider.

TABLE 1.5 Share of Discretionary and Nondiscretionary Spend by Category.

Particular	2013	2016	2021 (Predicted)
Private consumption expenditure INR billion	64,850	92,444	1,63,824
Nondiscretionary			
Food and grocery	22,503	30,691	50,294
Clothing	4799	6748	11,959
Total nondiscretionary	27,302	37,440	62,253
Discretionary			
Gross rent, fuel, and power	8560	12,295	21,952
Furniture, furnishing, appliances, and services	2594	3790	6881
Medical care and health services	2205	2681	3932
Transport and communication	10,571	14,051	22,608

TABLE 1.5 *(Continued)*

Particular	2013	2016	2021 (Predicted)
Recreation, education, and cultural services	1751	2404	3932
Food services	2477	3091	4979
Miscellaneous goods and services	9391	16,692	37,288
Total discretionary	37,548	55,004	1,01,571

Source: NRAI Report, 2016.

1.3.2 FOOD AND BEVERAGE SERVICE SEGMENTS AND MARKET SHARE ANALYSIS

To build an understanding of where the food service market currently stands in India and where it is headed, it is critical to analyze how the segments are playing out on the ground in terms of the market share of the overall revenue pie. The most reliable data for the above-mentioned purpose come from the National Restaurant Association of India. The below-mentioned table provides revenue data on all the key segments for the year 2016.

TABLE 1.6 Segment Wise revenue market share spread.

Format	Chain market size 2016 (INR crore)	Organized standalone 2016 (INR Crore)
Quick service restaurants (QSRs)	9125	9290
Affordable casual dining restaurants (ACDRs)	5110	36,820
Premium casual dining restaurants (PCDRs)	1610	6790
Café	1805	5370
Frozen desserts/ice cream (FD/IC)	1155	1690
Pubs/bar/café lounges (PBCLs)	1065	10,480
Fine dining restaurants (FDRs)	530	1815
Total market	20,400	72,255

Source: NHRAI Report, 2016.

1.3.3 CHAIN BASED VS STANDALONE: ORGANIZED FOOD SERVICE SEGMENTS AND MARKET SHARES

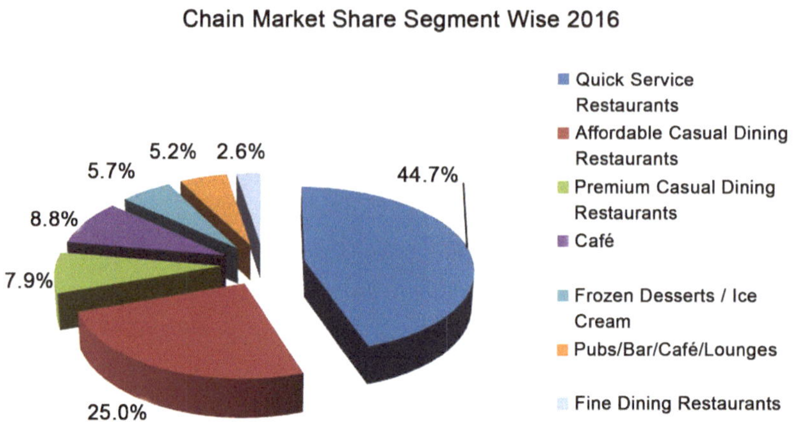

Interesting observations can be made through analysis of the data cited in Table 1.6, with respect to the chain format food and beverage service outlets. The dominance of the quick service restaurant and affordable casual dining restaurants is quite apparent as they represent nearly 70% of the revenue for the chain format businesses. This has implications for future investments in the space, and is up for interpretation as of the cited seven subsegments two control more than two-thirds of the market share. It will be interesting to see how the remaining segments chart their growth given that the urban–rural implications and socio-cultural shifts are continuing to define and guide consumption patterns in an economy that is expected to remain on a 7% plus annual GDP growth rate in the immediate future. As the disposable incomes expand more the 'Fine Dining Restaurant' segment should see growth, with regulations around alcohol consumption linked to the political decision-making and even possible judicial overreach. The Pubs, Bars segment remains more of a metropolitan city limited phenomenon and it remains to be seen how the socio-cultural, religious aspects determine both the supply side and demand side decision-making in years to come.

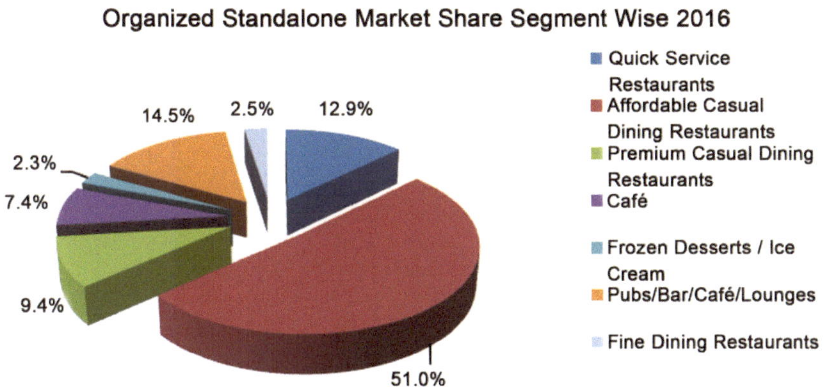

With respect to organized but nonchain standalone food and beverage service outlets, the affordable casual dining restaurant segment represents (as per the 2016 data cited in Table 1.6) more than 50% revenue market share. While pubs, bars, and cafes represented almost 15% of the revenue pie, interestingly, the QSR space was only at 13%, indicating the same to be more of a chain food and beverage segment phenomenon.

It is important to not lose the big picture here and also mention that revenue split between the chain and nonchain formats within the organized food service market continues to be dominated by the nonchain format at 77% of the revenue share with the chain-based segments sharing the remaining 23% as per the 2016 data shared earlier. The current status of the segments in terms of the respective market shares as well as the growth trends has the potential not only to guide just future investments, but also to enable efforts toward positive policy action depending on factors affecting their growth.

The table below offers some macro-level clarity on how the food and beverage service market has behaved from 2013 to 2016 and where it is headed in 2021 based on the trends that have been witnessed (Table 1.7) (NRAI Report, 2016).

TABLE 1.7 Indian Food and Beverage Service Market Share Trend.

Unorganized market	70%	67%	59%
The unorganized market share in the food services.	2013	2016	2021 (P)
Industry has reduced from 70% in 2013 to 67% in 2016 and is projected to fall to 59% in 2021. This is the case as many unorganized businesses are moving toward the organized sector. Policy action from the government also supports this shift. The pace of this change will depend on multiple factors but the direction is quite apparent.			
Organized standalone	22%	23%	28%
The organized standalone market share is projected to rise to 28% in 2021. This is the case as increasing disposable incomes have encouraged owners to open such organized outlets. Moreover, this segment projected CAGR of 15% from 2016 to 2021. The big impact will come from the tier 2 and 3 (cities) expansion.	2013	2016	2021 (P)
Chain market	5%	7%	10%
The chain market share increased from 5% in 2013 to 7% in 2016 and is projected to rise to 10% in 2021.	2013	2016	2021 (P)
Moreover, this segment has a projected CAGR of 22% from 2016 to 2021. The growth model of the chain format will be driven by franchise route for most part, venture capital funding, national as well as international will have play here.			
Restaurants in hotels	3%	3%	3%
The independent restaurants (hotels) market share is projected to remain constant at 3% till 2021. However, this segment is still growing with a projected CAGR of 9% from 2016 to 2021. How hotels respond to the competition outside for their customers is an interesting phenomenon to watch. they need to be innovative to ensure the food and beverage outlets inside remain viable and thrive. Many hotels have started considering outsourcing F&B operations, leasing spaces to outside players.	2013	2016	2021 (P)

Source: NRAI Report, 2016.

At a micro level, the subsegments essentially end up taking a cue from the macro-level shift from the unorganized to the organized format, which further includes both chain and nonchain-standalone operations. The below-cited data further establish the trends, CAGR (compounded average growth rate) charted by subsegments. What is quite apparent is that despite substantial investments and changes in consumption behavior led by economic expansion and varied other socio-cultural aspects there is huge potential for ongoing growth in both revenue terms and the number of outlets/operations.

In the standalone format of the organized market, the NRAI (2016) report (refer to Table 1.8) predicts that the "affordable casual dining restaurants" segment is likely to remain strong and grow to achieve a market share of 53% of the organized standalone space, with the quick service, premium casual dining, and the pubs/bars/café's segments far behind in terms of the revenue market share.

TABLE 1.8 Organized Standalone Market Share.

	CAGR 2013–2016	**CAGR (P) 2016–2021**	**Market share (2013)**	**Market share (2016)**	**Market share (P) 2021**
Café	6.9%	9%	8%	7%	6%
QSR	14.2%	15%	11%	13%	13%
PCDR	10.5%	20%	9%	9%	12%
ACDR	9.3%	15%	51%	51%	53%
FDR	3.5%	6%	3%	3%	2%
PBCL	7.6%	10%	15%	15%	12%
Frozen dessert/IC	8.5%	8%	2%	2%	2%

Source: NRAI Report, 2016.

In the organized chain format, the play seems to be consolidated between the quick service and the affordable casual dining. This is rather interesting as there seems to be a consumer preference toward branded chains when it comes to the above two segments, the same preference is not expressed when it comes to the remaining subsegments. The CAGR (predicted) for multiple chain segments, however, is strong as represented in Table 1.9 (NRAI Report, 2016) for the 2016–2021 period, which implies that the space is dynamic with strong growth fundamentals and potential.

How the food and beverage service market in India matures over the next couple of decades will be interesting to watch. If the last two decades have proven anything, it is likely to be one hell of a ride for the investors, operators, and of course the consumers.

TABLE 1.9 Organized Chain Market Share.

<table>
<tr><th></th><th>CAGR 2013–2016</th><th>CAGR (P) 2016–2021</th><th>Market share (2013)</th><th>Market share (2016)</th><th>Market share (P) 2021</th></tr>
<tr><td>Café</td><td>5.9%</td><td>10</td><td>12</td><td>9</td><td>6</td></tr>
<tr><td>QSR</td><td>18.4%</td><td>22</td><td>43</td><td>45</td><td>48</td></tr>
<tr><td>PCDR</td><td>19.4%</td><td>23</td><td rowspan="2">31</td><td>8</td><td>9</td></tr>
<tr><td>ACDR</td><td></td><td>20</td><td>25</td><td>25</td></tr>
<tr><td>FDR</td><td>1.9%</td><td>3</td><td>4</td><td>3</td><td>1</td></tr>
<tr><td>PBCL</td><td>25.4%</td><td>22</td><td>4</td><td>5</td><td>6</td></tr>
<tr><td>Frozen dessert/IC</td><td>14.1%</td><td>18</td><td>6</td><td>6</td><td>5</td></tr>
</table>

Source: NRAI Report, 2016.

1.4 CUISINES IN THE MIX

The Indian food and beverage services market, as may already be apparent, is a complex market to comprehend. The socio-cultural diversity, economic stratification, urban–rural divide, and role of religion in food consumption choices are all factors that impact how cuisines have received patronage and how they would find growth and acceptance in the future.

As globalization has swept much of the developing world, with the arrival of international food and beverage brands in the Indian market, the range of cuisine-related choices has only grown. The research documented by the NRAI in its 2016 report points out that the North Indian food still rules the roost at 28% market share followed by the Chinese at 19%. No surprise that most food and beverage service operations, especially in the standalone format, stick to the tried and tested formula of offering a multi-cuisine menu that essentially includes North Indian, Indian Chinese, and South Indian menu items.

TABLE 1.10 Share of Key Cuisines Across Formats.

North Indian	**28%**
South Indian	9%
Pizza	6.2%
Street food	6%
Regional Indian	3.1%
Pan Asian	2%
Chinese	19%
American	7%
Bakery and dessert	6%
Italian	4.5%
Continental	3.5%
Others	5.7%

Source: NRAI Report, 2016.

The above-cited data in Table 1.10 (NRAI Report, 2016), however, have to be viewed with a clarity that regional preferences for cuisines may not be in the same pecking order, nor would the rural–urban markets follow this break up. Depending on a targeted geographical market, chosen consumer segment a potential investor/operator should conduct a more focused research to understand the cuisine preferences.

With lifestyle-related diseases becoming rampant, urban India is witnessing a move toward consumption of organic food products. Vegetarian diet of varying levels of accommodation is driving chefs to look for nonmeat, nondairy menu items across the cuisines portfolio. Movements like "Tasting India" are generating interest in traditional food and ingredients aimed at preserving the culinary heritage of the country. A case in point is the revival and interest in millets (jowar, bajra, and raagi) etc., which were largely seen as the poor man's staple food (tasting India Symposium, 2018). Many a chef is recognizing the potential that the regional food offers and there is keen interest to give presence to regional as well as local street food on menus across the various subsegments of the food and beverage services sector.

Another focal point seems to be a renewed vigor with which sustainable culinary practices are being looked at, concepts like "slow food", "farm to fork", and more recently "plate to farm" are trending in the food and beverage service space in India. Munjal et al. (2016) offer a perspective on how food service businesses are recognizing the challenge of procuring farm fresh ingredients and are showing interest in back end integration, which is not all that simple, yet given the push toward sustainability. The restaurants are now attempting to grow their produce, develop herb gardens, or as a minimum forming alliances with local farmers along with the line of fair trade practices.

1.5 CONCLUSION

There are many more trends that are finding traction. The same will be discussed in detail later in the book. It will suffice to state that the opportunities that are waiting to be taken up, as the food and beverage services sector continues to grow and mature from a historically unorganized to organized form are immense. It is, however, equally true that despite the huge upside given the size of the market, financial success is challenging to put it mildly. The competition from both national and international investors and operators is immense as the customer seeks best in class services and products for the price and is loyal only to self.

As India continues to grow as an economic powerhouse, it is the services sector with its capacity to generate employment that will remain in focus; NRAI, Indian Food Services Report 2016 provides data on the direct employment generated by this segment of the services sector with a projection of nearly 9 million jobs by 2021 as shown in Table 1.11 (NRAI Report, 2016).

TABLE 1.11 Employment Share and Direct Employment (in Mn).

	Chain	Standalone	Unorganized	Total
2013	0.4	1.7	2.5	4.6 Mn
2016	0.6	2.3	2.9	5.8 Mn
2021 (P)	1.0	3.7	4.0	8.7 Mn

Source: NRAI Report, 2016.

The massive employment generation opportunity that the food and beverage services sector presents makes it ever relevant from a political standpoint. Initiatives like the skill India mission and setting up of sector specific "skills council" namely the "Tourism and Hospitality Sector Skills Council" are indicative of the policy response from the government aimed at supporting the growth story of the food and beverage service industry given the employability impact. Even as the governments at the center and state level fall over each other to move quickly to leverage the "tourism" opportunity essentially for the same reasons, the policy action in support of the food and beverage services segment still remains rather weak. In fact, anti-alcohol retail/ service legislations and judicial overreach as mentioned earlier have caused despair not just because such actions are seen as socio-culturally regressive, but also damaging to the profitability of key subsegments.

There is need for the food and beverage services sector at large to push its case for positive policy action such that the untapped potential is realized both from a perspective of employment generation, but more so to provide all key stakeholders including the investors, operators, employees, and customers the benefits aligned with their expectations.

KEYWORDS

- **food and beverage**
- **organized food retail**
- **unorganized food retail**
- **food service industry**
- **hotel food services**
- **restaurants**
- **F&B service industry**

REFERENCES

62nd Annual Report (2017–2018) *Federation of Hotel & Restaurants Association of India.*

Annual Report (2016–2017) *Ministry of Tourism*, Government of India.

Annual Report (2017–2018) *Ministry of Tourism*, Government of India.

Cooper, R.G. Winning at New Products. In *Accelerating the Process from Idea to Launch*, 3rd ed.; Perseus Books: New York, 2001.

De, D.; Munjal, S. Push for the food and beverage segment to drive revenue growth in Indian Hotels. In *The Indian Hospitality Industry: Dynamics and Future Trends*; Munjal, S., Bhushan, S., eds.; Springer: New York, 2017; Chapter 10.

Feltenstein, T. New-Product Development In Food Service: A Structured Approach. In *Cornell Hotel & Restaurant Administration Quarterly*, Vol. *27* (3), pp. 62–71.

Hospitality Trend Report. *AF and Company Consulting IBEF.ORG, India Brand Equity Foundation* [Online], 2018. https://www.ibef.org/download/Tourism-and-Hospitality-January-2018.pdf (accessed Apr 20, 2018).

IBEF.ORG. India Brand Equity Foundation. [Online] 2017. https://www.ibef.org/download/Tourism-and-Hospitality-January-2017.pdf (accessed Apr 5, 2018).
India Food Services Report. Published by National Restaurant Association of India, 2016.
Indian Hotel Industry. *Weaker than Expected Supply Addition during 2014-15 supporting Occupancies*, ICRA Research Services, 2015.
Indian Hotel Industry Survey. Published by Federation of Hotel and Restaurant Association of India in consultation with Hotelivate, 2017.
India Hospitality Industry Review. [Online] 2018, Published by HVS- Anarock, https://www.hvs.com/article/8488-hvs-anarock-india-hospitality-industry-review-2018 (accessed Apr 8, 2019).
Indian Hotel Industry Survey. Published by Federation of Hotel and Restaurant Association of India in consultation with Hotelivate, 2018.
JLL, India Hospitality Review. [Online] 2017. http://www.jll.co.in/india/en-gb/Research/Ready-steady-go-India-hospitality-industry-review-2017.pdf (accessed May 10, 2018).
Livemint.com. Economic Challenges for Modi 2.0. [Online] 2019, https://www.livemint.com/news/india/after-the-verdict-come-the-big-challenges-1558967086648.html (accessed June 13, 2019).
Munjal, S.; Bhushan, S. The Indian Hospitality Industry: Dynamics and Future Trends, Apple Academic Press: New Jersey, USA, 2017.
Munjal, S.; Sharma, S. Applying Innovative Food Cost Management Practices In Inflationary Times: Indian Budget Restaurant Segment Experiences. Worldwide Hospital. Tourism Theme. **2012,** *4* (5), 463–477. DOI: https://doi.org/10.1108/17554211211277897.
Munjal, S.; Sharma, S.; Menon, P. Moving towards "Slow Food", the new frontier of culinary innovation in India: The Vedatya experience. *Worldwide Hospital. Tourism Theme*. **2016,** *8* (4), 444–460. DOI: https://doi.org/10.1108/WHATT-04-2016-0022
Munjal, S.; Tiwari, S. Skill India Initiative – Implications For The Hospitality Sector In India. *Worldwide Hospital. Tourism Theme.* **2019,** *11* (1), 4–9. DOI: https://doi.org/10.1108/ WHATT-10-2018-0062.
Ottenbacher, M. C. Innovation Management In The Hospitality Industry: Different Strategies For Achieving Success. *J. Hospital. Tourism Res.* **2007,** *31* (4), 431–454.
Prakash, O. *Cultural History of India*. New Age International Publishers: New Delhi, 2004.
Tiwari, S.; Munjal, S. What are the key challenges that the Indian hospitality industry is facing in search of skilled labour? Worldwide Hospital. Tourism Theme. **2019,** *11* (1), 99–102. DOI: https://doi.org/10.1108/WHATT-10-2018-0063.
Sharma, S. Gastronomic Tourism in Old Delhi, CONTEXT: Built, Living and Natural, Volume 9, Issue 2, Autumn-Winter 2012.
Singh, A.; Munjal, S. Issues and concerns in technology decisions: the hotel operator's perspective. *Worldwide Hospital. Tourism Theme.* **2012,** *4*(2), 174–184. DOI: https://doi.org/10.1108/17554211211217343.
Tasting India Symposium, Swaad [Online] 2018. http://www.tastingindiasymposium.com/wp-content/uploads/2019/01/SWAAD-International-Day-of-Indian-Gastronomy.pdf (accessed Dec 18, 2018).
The Financial Express *Foreign Tourist Arrivals Grow By 4.4 Pc In 2015*, [online] 2016. http://www.financialexpress.com/article/economy/foreign-tourist-arrivals-grow-by-4-4-pc-in-2015/190929/ (accessed May 13, 2017).

CHAPTER 2

Toward Maturity: From Unorganized to Organized

ABSTRACT

The chapter provides a snapshot of sorts with respect to where the food and beverage service industry in India stands currently and where it is headed. The food and beverage services space has largely been operating in an organized format and it was only over the last two decades post the economic liberalization that the organized retail formats both in chain as well as non chain formats have found presence and growth. The chapter on one hand not only establishes the dilution in the market share of the unorganized part of the food and beverage business, indicating that growth of the organized segments will continue, but also argues the value and significance of the traditional businesses. The chapter also presents some interesting as well as informative statistics on the organized food and beverage formats with analysis that indicates key trends to look out for.

2.1 UNORGANIZED-ORGANIZED STATUS

The growth story of Indian economy from the first attempt at its liberalization in the early 1990s has been witness to a shift from an unorganized to an organized status; this has been true of many sectors and industries. Without getting lost in jargon, it will suffice to say that this shift is caused by complex economic forces getting to work in a manner such that there is consolidation, standardization, internationalization, and globalization and all the above are symptomatic of a maturing sector. Developed economies are quintessentially devoid of any significant presence of the unorganized format in any sector or industry in the economy.

India as a developing economy has struggled with this shift. The journey of the retail sector is one such example. After almost two decades of the liberalization of the economy nearly 85% or more of the retail market is still in the unorganized domain. The businesses in the form of mom and pop shops, "kirana" stores and similar in millions exist and control the earlier mentioned 85% market share. Despite the current status it is clear that the push by the organized players (Indian and international) will continue to shift this market share mix such that the unorganized format will continue to loosen its grip. The courier and logistics sector too witnessed a similar journey of consolidation and shift from largely localized unorganized businesses running on the ground to now mostly national, organized players with significant market share dominating the market. Will the food and beverage service sector behave any differently is a good question to ask, even if an answer is not all that forthcoming as yet.

The unorganized food and beverage service market share in 2016 was 70% and as per the NHAI 2016 Report expected to get diluted to 59% by 2021. The organized formats including the chain-based formats, organized stand-alone, and restaurants in hotels split the pie rather unevenly with the stand-alone segment dominating. Hotel-based food and beverage service outlets interestingly have a rather insignificant market share at about 3%, which is not expected to change much. Chapter 1 has already cited and discussed the market shares. The larger issue to discuss here is not just the trend but understanding its implications for investors, operators, and consumers. FICCI Report (2015) also recognizes both the immense potential as well as the shift that the food and beverage service sector in India has experienced. While there seems to be a consensus on the overall market share dilution in the unorganized space, there are minor differences in how the 40% odd market share of the organized segment will be split; though again there is no debate that the stand-alone formats will hold sway over the organized chain formats in the time to come.

2.1.1 FROM NUKKAD CHAIWALA'S TO MODERN F&B OUTLETS AND SUCH

Generations of Indians grew up enjoying tea and a savory snack or two at the local "chaiwala" (tea vendor) at a street corner. While rural India may have bucked the trend, this constant in the lives of millions in metropolitan cities is fast disappearing as more trendy, often air conditioned; cafes have come in place, providing comfort and service that the next generation of consumers expects and is willing to pay for. Nostalgia around the "nukkad" tea stall notwithstanding, the superior ambiance, comfort, service standard, hygiene, and standardized product (not to mention the expanded range of product offerings) is here to stay and eventually find way into smaller towns across the length and breadth of the country. One can debate the gradual loss of the traditional chai shops or joints where the quality of the product was a result of the inimitable effort of the vendor, often known far and wide for his special way of making tea, but there is no going back as generation Z takes over. FICCI Report (2015) is categorical in stating that the youth of India by virtue of their sheer numbers as well as eating habits will define the shape of the food and beverage services sector.

FIGURE 2.1 Tea and snacks the old way on the street.

Source: Author.

The same holds true for the nondescript small time family run restaurants, local eateries or sweet shops that were the only potential eat out options available, usually serving regional/local cuisines with rather basic seating arrangements and typically no air-conditioning. As sociocultural change brought in more nuclear families, with more disposable incomes, eating out became a trend in the urban cities across the nation. The consumer expectations for more services, higher standards in terms of the ambiance and products have resulted in the organized formats (chain or stand-alone) find traction. Issues like poor hygiene and lack of basic infrastructure has pushed the traditional eateries to either invest and evolve into organized food and beverage service setups or close doors.

FIGURE 2.2 Unorganized format restaurant in Old Delhi.

The rural India and the small/medium towns across the nation still have the traditional unorganized format thriving but it is quite evident that over a period of time these places too will see more outlets akin to the organized formats finding presence and growth. Depending on what is one's perception and understanding of development and economic growth, this imminent change in the landscape of commercial food and beverage services can be subjected to varying interpretations. At some level the traditional tea stalls, dhabas, sweet shops, and local eateries being replaced by modern day dinning formats is a loss of our cultural heritage and also an economic loss in the form of livelihoods that these small businesses, unorganized as they may be, are being subjected to. Will there be any space for these traditional eating formats? Will they be able to adapt and survive the push from the organized formats of food and beverage services? These questions will find their own answers over the next few years. The growth of organized retail over the last two decades impacted the small shop owners, often identified as the "kirana shops" but they too innovated, adapted and used their low overheads and personalized services as a competitive advantage, thus, giving the larger organized and branded formats a run for their money. This parallel of sorts can be a reference point for the unorganized food and beverage segment in India. That said the arrival of the online shopping vertical has added another dimension for the retail trade to grapple with, questioning the long term viability of the "brick and mortar" store format with the fast growing "virtual store" format finding traction. If media reports are to be believed (The Hindu, 2019), the organized retail should double in size by 2021 and continue to grow.

2.1.2 DOES TRADITIONAL CULINARY EXPRESSION HAVE TAKERS?

Given how dynamic the business environment is, technology-related change is likely to remain a force in how businesses will grow and/or survive; food and beverage service businesses will be no different. That said there is a parallel push toward recognizing the need to protect, conserve, and rehabilitate traditional knowledge systems and culinary heritage. Slow food India, Tasting India Symposium and many more such platforms are working toward providing voice and support to the indigenous food and beverage service providers, building momentum to bring in place policy action and decision-making at highest levels of the government so as to ensure that unorganized formats are recognized for the cultural value they represent, and build public sentiment toward ensuring that these businesses receive the patronage required to keep them in business and ensure they thrive. The response toward protecting our culinary heritage, use of ethnic ingredients and following traditional methods of cooking in various communities is gaining interest; this bodes well for traditional Indian food and beverage service points of sale (The Economic Times, 2015).

The shift for traditional, old school unorganized food and beverage businesses in India to an organized stand-alone format is not all that drastic a move, many stand-alone outlets/restaurants have a legacy that takes them back to times when the nation got independence from British rule in 1947, Essex Farms, Moti Mahal and Kwality Restaurant in Delhi, Leopold Cafe and The Table in Mumbai, Vidharthi Bhawan and Koshy's in Bengaluru and Glenary's in Darjeeling (Fig. 2.3) are examples of many such names that generations grew up patronizing. These stand-alones are inherently managed by families, passed on from generation to generation. Since the entire gamut of operations is managed by the ownership directly, they have a distinct character that is represented in both the product and the services being provided. These have best of tradition and professionalism to offer.

FIGURE 2.3 Family run stand-alone restaurant in Darjeeling.

Source: Author.

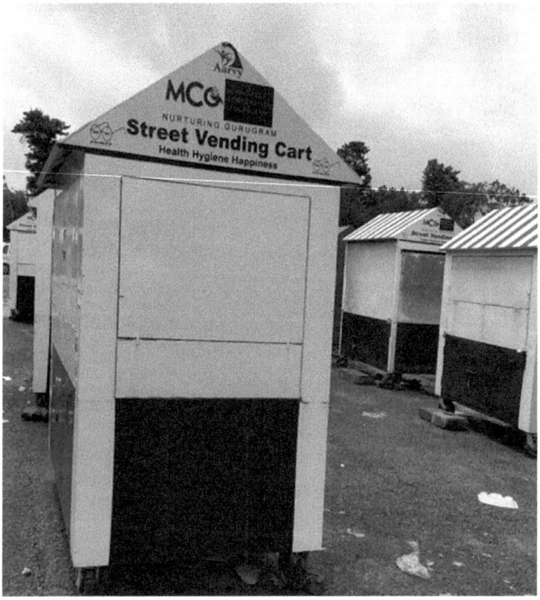

FIGURE 2.4 Street vending cart—a government initiative.

Source: Author.

As a more recent trend many a stand-alone has come up as an entrepreneurial initiative of young chefs who wanted to establish their own signature brands. Chef Abhijeet Saha (Caperberry), Ritu Dalmia (Diva), Marut Sikka (Kainoosh), and Pooja Dhingra (Le 15 Patisserie) are a few prominent examples of restaurant brands led by the respective chefs. As mentioned earlier, some stand-alone food and beverage businesses have also chosen to expand by adding units resulting their taking the shape of regional or national chains based on their geographical presence and size. Cafe Delhi Heights, Yum Yum cha, Insta Pizza, and Keventers are a few examples in this respect. Some of the prominent Café chain brands and their outlet strength as of 2016 is shared in the table below. Café format seems to be one of choice in this regards.

TABLE 2.1 Prominent Cafe Chain Brands and the Number of Outlets.

Key brands	Number of outlets
Cafe Coffee Day	1556
Monginis	409
Barista	141
Just Bake	116
Starbucks	81
McCafe	75
The Chocolate Room	69
French Loaf	68
Costa Coffee	47
Chai Point	45
Mad Over Donuts	43
Birdy's	39
The Coffee Bean & Tea Leaf	33
Krispy Kreme	31
Chaayos	25
Daily Bread	22
Tea Trails	19
Au Bon Pain	18
Di Bella Café	15
Mocha	12

Source: NRAI Report, 2016.

Organized stand-alone restaurants need to compete with chain restaurants. This competition is more intense in certain subsegments than others, and this will become apparent as we analyze available data. Chain formats are the obvious choice for international brands as the intent to expand rapidly to achieve national market presence and economy of scale is quite clear. The chain formats have the resources to use mass media marketing approach and tend to centralize certain aspects of operations like purchase and even product development. The focus is on ensuring a standardized product-service model that is delivered in a cost effective way. The stand-alones rely on digital media more than others to market themselves. They usually have shoe string budget to work with and thus find the digital media more conducive. Along with the traditional word of mouth approach, personalization and value creation through constant innovation is what drives them and enables them to attract as well as retain customers. Table 2.2 below offers perspective on the positioning, target market, location, and other aspects for the specific subsegments (or formats) of chain-based operations in the Indian context.

TABLE 2.2 Food and Beverage Services Chain Market—Format Description.

Type of chain	Positioning	Target group/ pricing	Ambience	Locations
Cafe	Tea–coffee centric with limited focus on food	15–45 years/ affordable	Relaxing, "catching up" and unwinding	Malls, high streets, popular markets, office complexes, airports, hospitals, highways, educational campuses
QSRs	Specific product offerings or cuisines with focus on convenience	15–35 years/ affordable	Functional interiors, compact seating with self service	Malls, high streets, popular markets, office complexes, airports, hospitals, highways, educational campuses, multiplexes

TABLE 2.2 *(Continued)*

Type of chain	Positioning	Target group/ pricing	Ambience	Locations
ACDRs	Focus on moderately priced food and table service	20–50 years/ affordable	Casual fun environment. Regular seating with table service	Malls, high streets, popular markets, office complexes, airports, hotels
PCDRs	Focus on quality, presentation and service	25–50 years/ casual to premium	Up market environment with design-led ambience	Malls, high streets, popular markets, office, complexes, food hubs
FDRs	Multi-cuisine/ specialty restaurants with a focus on quality, ingredients, presentation and service	25–50 years/ premium to luxury	Formal, premium design-led ambience	Malls, high streets, popular markets, office complexes, hotels
PBCL	Focus on alcohol and customer experience	20–40 years affordable to premium	Theme based with music, rustic, no frills to classy	Malls, high streets, popular markets, office complexes, airports, hotels
FD/ICs	Focus on ice creams and frozen yoghurt with limited options for snacks/ beverages	18–35 years/ affordable	Colorful, fun, focus on takeaways	Malls, high streets, popular markets, office complexes, airports, hospitals, highways, educational campuses, multiplexes

Source: NRAI Report, 2016.

2.2 ORGANIZED FOOD SERVICE MARKET AND ITS SUBSEGMENTS AT A GLANCE

The food and beverage services market in India was estimated to be around INR 3,09,110 crores in revenue terms in 2016 and expected to be growing at CAGR of 10% with a forecast to be worth INR. 4,98,130 crores in 2021. The organized market in 2016 was estimated to be nearly INR. 1,00,000 crores with a forecast to reach INR. 2,10,000 crores by 2012. There is limited creditable data to dissect the unorganized segment; however, the NRAI (2016) Report has done a good job to share some valuable data on the subsegments of the organized part of the Indian food and beverage services sector. The FICCI Report (2015) largely concurs with the broad estimates. The section below analyses the market share split in revenue terms between the ***organized stand-alone*** and **organized chain formats** for each of the seven classified subsegments or F&B service formats.

a. **Quick Service Restaurants** (QSR) as a segment of the organized food and beverage service market seems to be evenly split between the stand-alone and chain formats. Both national and international brands have significant presence in the Indian market. Established stand-alone restaurants have expanded into chain formats and the trend is likely to continue. The segment owes its success and expansion to multiple factors. The key aspect is the changing

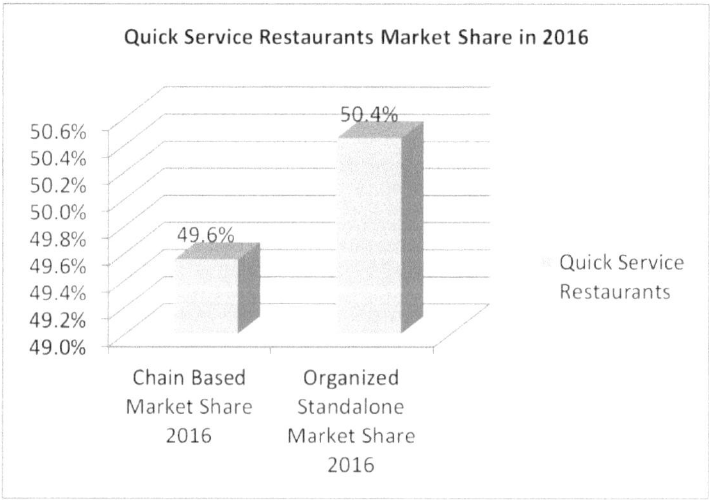

sociocultural context. As a forward path, the next wave of growth is likely to come from the tier 2 cities. The chain market is fairly consolidated as top five players control nearly 77% of the market with Domino's taking a major share at 30%. The chain format has nearly 110 brands with 3600 plus outlets. The level of consolidation currently seen is likely to shift as the semi urban India comes into play. Currently, 70% of the market is limited to metro and mini metro cities. Tier 1 and 2 cities account for the remaining 30%. The regional spread is rather even with South India having 34% of QSR outlets, North and West having 30% each.

TABLE 2.3 QSR Food and Beverage Service Segment Data (2016).

Chain Market Share by Number of Outlets	
Dominos	30%
Subway	16%
McDonald's	11%
KFC	10%
Goli Vada Pav No. 1	10%
Others	23%
Region-wise Outlet Presence	
North	29%
South	34%
East	7%
West	30%
City-Type Outlet Presence	
Metro	34%
Mini Metro	35%
Tier I	12%
Tier II	19%
Dine In and Takeaway/Delivery Mix	
Dine in	59%
Takeaway	9%
Delivery	32%

Source: NHAI Report, 2016.

Indian QSR brands like Goli Vada Pao, Bikanerwala, and Haldiram's have vegetarian offerings, while the international brands have a mix of both vegetarian and nonvegetarian menu options. QSR space in India is food centric, both in terms of SKU mix as well as revenue share. Nearly 80% of QSR business is about food. From a perspective of sales mix between dine in and takeaway/delivery options, the preference for dine in at nearly 60% currently is likely to shift as brands strengthen their delivery operations and infrastructure.QSR space in India is expected to be worth INR 24,500 crores by 2021 (NRAI, 2016).

b. **Café** culture is where the organized food and beverage service segment may have found its initial momentum. As a tea crazy nation with a growing economy, changing sociocultural dynamics and associated demand side implications sought branded beverages in superior service setting (read air-conditioned, well appointed interiors). The arrival of Café Coffee Day, Barista, and many others set the tone for stand-alones to take the cue. Soon enough the metros saw middle class Indian's thronging the cafes to meet friends, discuss business, meet for arranged marriage interactions, and many other social settings. The fact that chain-based outlets have only 25% market share in revenue terms implies how the concept struck a chord with the stand-alones and resulted in them taking two thirds of the market share.

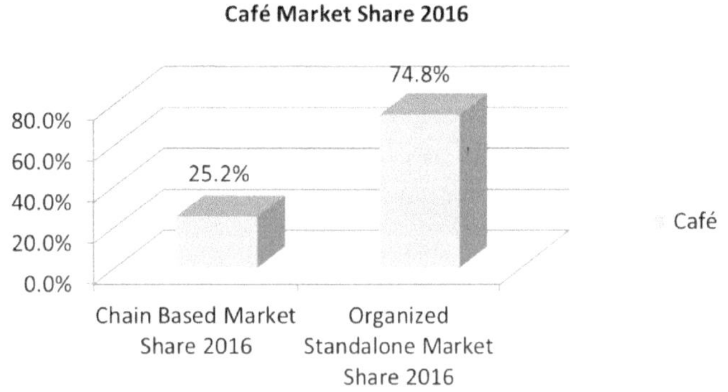

Another interesting facet of this segment's growth has been the new found interest in Coffee as a beverage in a largely tea focused market historically. While developed nations have a per capita consumption of Coffee at 8 kg a year. In India, in stark comparison it is still less than 120

gm a year. As the trend finds presence in the semi urban, tier 2, 3 cities these statistics are likely to shift upwards. The café market worth INR 1805 crores in 2016 is likely to be worth INR 2905 crores by 2021 (NRAI, 2016). The café market is further segmented into premium (Starbucks, The Square) brands and the regular cafes. The competition is intense as both international as well as domestic brands find play, both chain and stand-alone formats remain vibrant. Some prominent new players include The Chocolate Room, Chaayos, Chai Point, Wagh Bakri, and Infinitea to name a few. The resurgence of tea as a long standing beverage of choice is quite apparent. Interestingly cold beverages are being as well received as the hot peers. Keventer's has reinvented itself with a strong push for premium milkshakes and much more. The café market currently has more than 100 chain brands with 3000 plus outlets; stand-alones are in much larger numbers, 12000 and growing. Café Coffee day with 50% market share of chain-based format is the market leader in a segment where innovation in terms of product-service combine is a way of life. The segment is finding growth as it reaches out to the Indian middle class with a growing spending capacity, even hospitals. Office set ups are potential café locations as the format continues to reach its customer base. The table below summarizes the retail formats and inputs on space dynamics and kitchen infrastructure. The café market that was worth about INR 1800 crores in 2016 is expected to grow to INR 2900 crores by 2021 (NRAI, 2016).

TABLE 2.4 Cafe Retail Formats.

	Lounge	**Cafe**	**Kiosk**
About the format	Premium format with emphasis on ambience along with F&B experience	Format with comfortable seating area	Smallest format in terms of size with limited or no seating
Space dynamics	Elaborate cafe with premium comfortable seating	Smaller than lounge but similar number of seating	Smallest amongst all the formats, with no seating
Food preparation	Usually includes kitchen where some items are cooked and prepared fresh	Pantry area used to freshly prepare beverages and heat/ grill other items	Beverages prepared freshly limited precooked/packaged food
Representative brands	CCD Lounge & Square, Mocha	Starbucks, CCD, Barista, Costa Coffee, Coffee Bean & Tea Leaf, Au Bon Pain	CCD, Barista, CBTL, Costa Coffee

Source: NHAI Report, 2016.

The SKU mix for the café segment remains equally balanced between beverages (both hot and cold) and food. Food includes vegetarian as well a non-vegetarian options. Interestingly street food, regional snacks, bakery products are finding presence on the café menus as they try and compete on food variations and menu variety. Chain formats have nearly 80% of the outlets situated in Metro, Mini Metro cities; Western region leads with 45% share of the number of outlets, with South and North following at 29% and 20% respectively. The Café space is largely a Dine-in play with 75% sales; it will be interesting to see how takeaway option performs in time to come. Café segment of the food and beverage services market is interesting as it is not comparatively capital expenditure heavy as compared to other segments; as a result entrepreneurship opportunity in the form of standalone format thrives and is likely to remain a growth segment NRAI (2016).

TABLE 2.5 Cafe Chain Brands and the Number of Outlets.

Key brands	Number of outlets
Cafe Coffee Day	1556
Monginis	409
Barista	141
Just Bake	116
Starbucks	81
McCafe	75
The Chocolate Room	69
French Loaf	68
Costa Coffee	47
Chai Point	45
Mad Over Donuts	43
Birdy's	39
The Coffee Bean & Tea Leaf	33
Krispy Kreme	31
Chaayos	55
Daily Bread	22
Tea Trails	19
Au Bon Pain	18
Di Bella Café	15
Mocha	12

Source: NRAI Report, 2016.

c. **Casual Dining Segment** is all about attracting India's massive middle class, especially in the metro, Tier 1–3 cities that are interested in a dining out experience but have a budget to work with and also looking at food options that are different from what QSRs have to offer. The segment can be further broken down into affordable casual dining and premium casual dining. Basic service, good neat clean ambiance (air-conditioned or not), and menu that has something for all age groups is what works with the affordable casual dining market segment. Another quite interesting facet is the lack of chain-based organized food service outlets present. As on 2016, nearly 88% of the market share was retained by the stand-alones. If one reads a little more into this data, it looks like the unorganized segment of the food and beverage service market is transitioning into this space by making

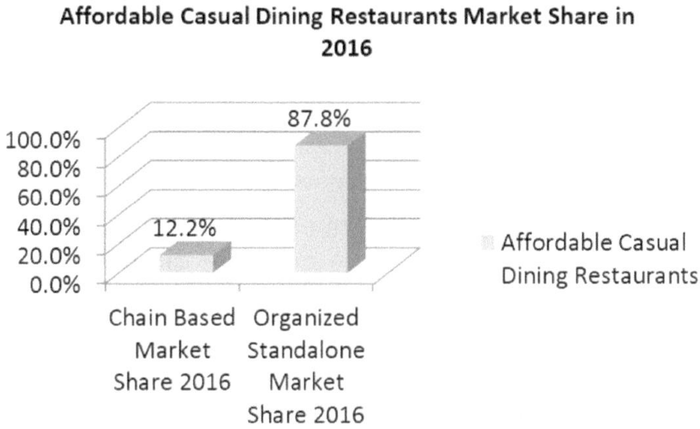

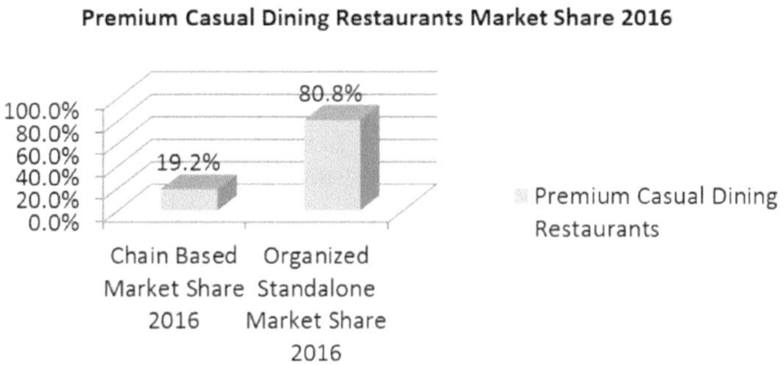

an effort to improve services and infrastructure and other markers that separate the unorganized from the organized segment. If this analysis is sound it will be safe to suggest that as the shift from unorganized to organized formats continues this subsegment is going to continue to grow as more and more traditional restaurants transition to the stand-alone format in the years to come. The premium casual dining segment too seems to be more of a stand-alone format space but that seems to be changing as stand-alone start ups are now growing into chains. Farzi Café, Soda Bottle Opener Wala, Pind Balluchi, and Sagar Ratna are good examples to cite for quick reference. While the space is largely dominated by domestic brands, Pizza Hut with a 21% market share of the chain format of casual dining segment is worth mentioning. Nearly 50% market is concentrated with businesses having 3–4 outlets. There is also movement of players from the QSR space venturing into the casual dining market given that such transitions can be made with some subtle changes in service–product offerings. The casual dining restaurant (CDR) segment has about 1400 outlets spread over the country. Nearly 80% of these are based out of metro and mini metro cities while remaining find presence in tier 1 and 2 cities. In sharp contrast with earlier discussed segments, half of the outlets are in northern region, while the other half is divided between south and west regions. The casual dining segment is all about food, be it in terms of sales or SKU mix, food constitutes 75% of the revenue share. Dine in option has a 60% food component. Same pattern is evidenced by the takeaway/delivery formats. The casual dining market is estimated to be more than INR 6720 crores as per the NRAI Report (2016).

d. **Fine dining** restaurant space is capital intensive. For many a decade, fine dining restaurants by definition were confined to 5 star luxury hotels situated in the metro cities. There were few stand-alones to choose from as well but largely the space was confined to luxury hotels. Socioeconomic factors as well as an overlaying cultural dimension to eating out ensured that a very small part of the population had the means to experience fine dining at these restaurants. In many a case, dining at these restaurant was restricted to very special life occasions like anniversaries, birthdays, or marking other special personal/career achievements. Being able to afford such dining experiences was symbolic of one's social position and affluence.

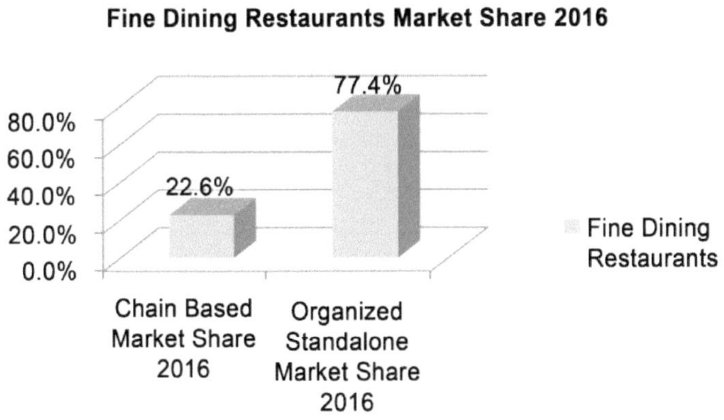

True to name, the fine dining restaurant segment is all about food made from high-quality ingredients, well presented with excellent service. The fine dining is nothing but a packaged experience that is customized/ personalized to the needs of the customer. This includes the ambiance and setting, the food, and finally the service. Today, one can find FDR space expanding, offering Indian and international cuisines, authentic food and experience to the patrons. A review of the market share of the segment indicates this segment too has a strong play for the organized stand-alone format as compared to the chain format. Table 2.6 below provides more data on the fine dining segment which is estimated to be about INR 530 crores, and predicted to grow to about INR 615 crores by 2021, NRAI (2016).

TABLE 2.6 Fine Dining Restaurants.

Prominent Chains Market Share by Number of Outlets

The Great Kebab Factory	28%
Punjab Grill	22%
Yautacha	10%
Royal China	8%
Others	32%

City-Type Outlet Presence

Metro	65%
Mini Metro	30%
Tier I	4%
Tier II	1%

Zone-Wise Outlet Presence

North	36%
South	25%
East	4%
West	35%

Dine In and Takeaway/Delivery Mix

Dine in	91%
Takeaway	3%
Delivery	6%

Sales and SKU Mix

Sales		SKU	
Food	68%	Food	40%
Beverage	25%	Beverage	56%
Dessert	7%	Dessert	4%

Sales Mix—Business Hours

Hours	**Weekdays**	**Weekends**
2200–0000	13%	14%
1900–2200	53%	32%
1600–1900	1%	2%
1300–1600	31%	50%
1100–1300	2%	2%

Source: NRAI Report, 2016.

e. **Pubs, Bar Cafés, and Lounges** is a food and beverage service segment that is all about beverages (alcoholic as well as nonalcoholic). It represents a sociocultural shift that defines urban India as it pursues the lifestyle typical of developed western economies. It is one of the fastest growing segments with a CAGR of 22% till 2021. The space is not only witnessing constant innovation both at the level of product and service models, but also in terms of overall concepts with which investors are making an effort to get consumers interested. Café's with a bar license to sell and serve alcoholic beverages are being well received, so are the pubs focused on serving fresh beer to patrons along with a range of alcoholic beverages. The segment had a market size of nearly INR 1100 crores in 2016 and is expected to grow to nearly INR 3000 crores by 2021. The market is largely a organized stand-alone format play; however, some popular chain brands are The Beer Café, TGIF, Pop Tate's, 10 Downing Street, Hard Rock Café, and Bar Stock Exchange to name a few. The segment is also leveraging growth of professional sport in India, as cricket, football, kabaddi, and other sports develop commercial formats/leagues etc. The pubs, beer café's and lounges (PBCL) segment is found promoting these events to build their customer base. The segment, however, is quite fragmented and competitive as well.

Pubs/ Bar Café /Lounges Market Share 2016

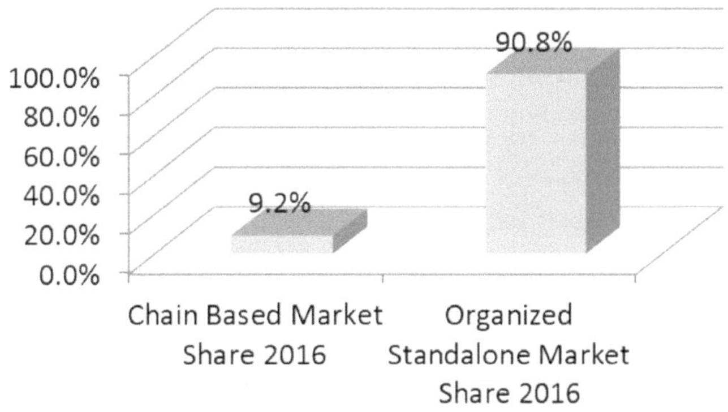

In terms of outlet presence, more than 90% of the outlets are located in the metro, mini metro cities. Given that some states are dry (Gujrat, Bihar, and Tripura are examples) and hence out of bounds for selling alcohol/ alcoholic beverages, the regional spread is affected accordingly. Western region at 43% leads the way, followed by north at 32% and south at 23%. Dine in is the delivery format of choice.

f. **Frozen desserts, ice cream, and milkshakes** in a branded format are a developing market in India. The strengthening of the cold chain (supply side) along with a social acceptance around eating out is possibly the main reason for the segment to gain traction. India's per capita consumption of ice cream is at 400 mL as compared with a global average of 2.3 L. That said the segment which was worth about INR 1200 crores in 2016 is expected to be worth INR 2640 crores by 2021, growing at a CAGR of 18%. Both chain-based as well as stand-alone formats have significant play. Though stand-alones at 60% market share have a larger presence. The market has evolved from a largely unorganized, mom and pop shop store-based, mobile cart-driven segment to one where parlors and well-appointed kiosks are offering a wide variety of ice creams, frozen desserts, and more to their patrons. Retail brands like Mother Dairy, Kwality Walls, and Amul are still around in a new avatar; however, many international brands like Baskin-Robbins, Haagen-Dazs, Swirls, Naturals, and Havmor are building presence in the premium parlor market segment. These brands are constantly innovating to offer new toppings and flavors. Sugar-free or low-fat variants for the health conscious or diabetic customers are also being made available. Experimentation in the form of ice cream sandwiches, pita pockets, and more is also indicative of how dynamic the space is in terms of product offerings.

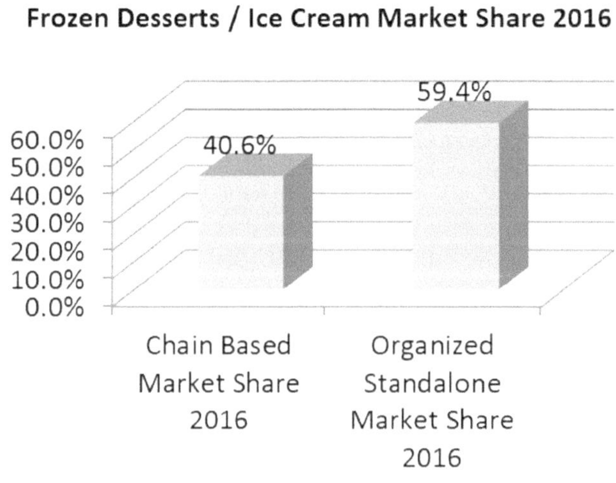

> Keventer's has energized the milkshake space within this segment as the brand now boasts of more than 200 kiosks/parlors pan India. In the premium ice cream segment, Baskin Robbins with a market share of 41% leads the pack.

With about 40 odd brands and 1700 plus outlets pan India, the segment is on its way. The regional spread of the market currently has western region leading at a 39% market share in terms of number of outlets, followed by south and north regions at 28% each. Metros and mini metros account for 65% of the outlets while the remaining is spread over tier 1 and 2 cities across the country. In terms of SKU mix and sales, desserts account for nearly 90% share, though beverages are also being offered by most brands. It is not surprising that the segment has to deal with the seasonality aspect, the March–June period accounts for 50% of the sales. There is, however, a regional twist to seasonality as weather patterns are different for each of the four geographic regions NRAI (2016).

The game changer for this segment has been the significant improvement in the supply side back end, especially the cold chain. Given the overall size of the Indian market, this is just the beginning as development of this organized F&B format is likely to continue and evolve.

2.3 FOOD AND BEVERAGE SERVICES MARKET—OPPORTUNITIES AND CHALLENGES

The journey of the food and beverage services space over the last decade or so makes it apparent that there is immense growth potential that remains to be tapped. The shift from an unorganized to organized format (chain/stand-alone) has been journey that has seen emergence of multiple subsegments. These have been discussed in this chapter briefly in order to provide a better perspective. It is the growing Indian middle class with spending power and a will to experiment with food that makes this space exciting for not just operators and investors but hospitality professionals as well. The road ahead will be bumpy as there is intense competition for market share with need for constant innovation that is to be expected as well.

The segments will grow and find acceptance in the tier 1 and 2 cities, the market will mature and evolve with time. Table 2.7 below provides a consolidated perspective on the opportunities as well as challenges that the businesses are likely to encounter. In addition to the format-specific challenges, high cost of real estate, skilled manpower, and over regulation as some common challenges that the industry faces, these will be discussed detail in a subsequent chapter.

TABLE 2.7 Opportunities and Challenges for the Food and Beverage Service Formats in India.

Type of segment	Opportunities	Challenges
Cafe	• Increase in "hang out culture," a place for socializing • Scalable model in smaller format • Large play in food and beverage based innovation • Tier 2, 3 cities and semi urban India is up for grabs • Both chain and stand-alone formats are going to work	• Low margins • Limited brand salience and loyalty • High table turnaround time vis a vis per table spend
OSRs International Brands	• Increasing time poverty among youth • Popularity of international cuisines • Improved supply chain, cold chain • Lifestyle shift, sociocultural changes support need for quick service options with delivery, takeaway options	• Highly price-competitive market • High investment in equipment • Highly price-competitive market
OSRs—Indian Brands	• Increasing popularity of the format • Increasing focus on convenience • Menu that caters to the Indian masses • Stand-alone operations have an opportunity in this space, they can also grow	• Mostly family operated and lack of management bandwidth • Competition from street food • High tax implication vs nothing being paid by the unorganized segment
CDRs	• Increasing middle and upper middle class in urban India • Increasing incidence of eating out • Increasing value for money focus	• Fragmented market with increasing competition • Maintaining consistency across location
FDRs	• Rising disposable income • Increasing focus on premium lifestyle	• Shorter shelf life • Require constant innovation and differentiation • High fixed operating costs

TABLE 2.7 *(Continued)*

Type of segment	Opportunities	Challenges
PBCL	• Favorable Indian demographics with 65% of population below 35 years • Increasing incidence of alcohol consumption • Changing lifestyle	• Price competitive segment with competition from in-home consumption • Setting up process is cumbersome with extensive licensing process
FD/ICs	• Popularity across age segments with increasing spends on desserts • No more a seasonal business • Scalable model as they require smaller formats	• Low transaction value • Competition from FMCG products in FD/IC segment

Source: NRAI Report, 2016.

2.4 CONCLUSION

Food and beverage service space in India will grow in response to the changing sociocultural economic landscape in a rather interesting way. There are multiple factors driving this growth—pressure of nuclear families with both husband and wife working, higher disposable incomes, increasing acceptance of eating out as a way of life, and broadly changing lifestyles. These factors will impact the growth of both the unorganized as well as organized (chain and stand-alone) food and beverage service formats. The organized formats (especially chain based) are finding investment support both in the form of FDI and venture capital leading to a sharp growth in number of outlets spread over the metropolitan and tier 1 Indian cities. It is this expansion that has started to scale up resulting in a trend that will see unorganized format being reduced to a 60% overall market share. That said the changing sociocultural dynamics in rural and semi urban India are the next pit stop. Traditions, customs, and religious interlinks will have a profound bearing of what comes next for the F&B service industry. One can expect the change in the Indian hinterland to be slow as against the rather rapid shift that was witnessed in the urban centers.

Factors like internet access and cheap mobile data have opened up information sharing like never before. This alone seems to be driving sociocultural changes, generation Z from the rural and semi urban India will determine the next phase of expansion and growth of the various organized F&B service formats. As mentioned in the chapter earlier, some formats are likely to find interest while others may have to wait for a while before these markets open up to what they have to offer. "Bar cafes" or "pubs" finding takers in rural India can be one such example, same may be the fate of "fine dining restaurants"; however, some of the other formats like the "quick service restaurants" or even "CDRs" may find acceptance from the younger generation of Indians from rural, semi urban regions. Some of this growth would come from new operations being set up, while many are a result of existing food service business owners making a move to operate like organized players in the market (Food Service India, 2019).

Whatever may be the case, it sure is going to be an interesting journey, it will be critical for the businesses to recognize the sociocultural diversity of the land, how regions (south, north, east, and west) are so different, how religious aspects can be critical aspects to watch out for. And all of the above and more is what makes India a market like none other in the world.

KEYWORDS

- **food and beverage service**
- **fine dining**
- **QSR**
- **cafes**
- **pubs**
- **organized food retail**
- **casual dining**
- **Indian F&B market**

REFERENCES

62nd Annual Report (2017–2018) *Federation of Hotel & Restaurants Association of India.*

Annual Report (2016–2017) *Ministry of Tourism*, Government of India.

Annual Report (2017–2018) *Ministry of Tourism*, Government of India.
IBEF.ORG, India Brand Equity Foundation [Online] 2017. https://www.ibef.org/download/Tourism-and-Hospitality-January-2017.pdf (accessed April 5, 2018).
IBEF.ORG, India Brand Equity Foundation [Online] 2017. https://www.ibef.org/industry/retail-india.aspx (accessed January 4, 2019).
De, D; Munjal, S. 'Push for the Food and Beverage Segment to Drive Revenue Growth in Indian Hotels'. In *The Indian Hospitality Industry: Dynamics and Future Trends;* Munjal, S, Bhushan, S., Eds.; Springer: New York, 2017.
Food Service India, Key trends impacting food service industry in India [Online] 2019. https://www.indiaretailing.com/2019/01/17/food/food-service/expert-speak-key-trends-impacting-foodservice-industry-in-india/ (accessed May 10, 2019).
FICCI Report, Unlocking the Potential in Food and Beverage Services Sector [Online] 2015. http://ficci.in/spdocument/20589/F&B-Report.pdf (accessed June 12, 2018).
India Food Services Report 2016, Published by National Restaurant Association of India.
Indian Hotel Industry Survey 2017, Published by Federation of Hotel and Restaurant Association of India in consultation with Hotelivate.
India Hospitality Industry Review [Online] 2018. Published by HVS Anarock. https://www.hvs.com/article/8488-hvs-anarock-india-hospitality-industry-review-2018 (accessed April 8, 2019).
Indian Hotel Industry Survey 2018, Published by Federation of Hotel and Restaurant Association of India in consultation with Hotelivate.
Hospitality Trend Report, Published by AF and Company ConsultingIBEF.ORG, India Brand Equity Foundation [Online] 2018. https://www.ibef.org/download/Tourism-and-Hospitality-January-2018.pdf (accessed April 20, 2018).
JLL, India Hospitality Review [Online] 2017. http://www.jll.co.in/india/en-gb/Research/Ready-steady-go-India-hospitality-industry-review-2017.pdf (accessed May 10, 2018).
Munjal, S.; Bhushan, S. *'The Indian Hospitality Industry: Dynamics and Future Trends'*; Apple Academic Press: New Jersey, 2017.
Sharma, S. Gastronomic Tourism in Old Delhi, CONTEXT: Built, Living and Natural **2012,** *9* (2), Autumn-Winter 2012.
Sandeep, M.; Sanjay, S.; Menon, P. "Moving towards "Slow Food", the New Frontier of Culinary Innovation in India: The Vedatya Experience", Worldwide Hospitality and Tourism. *Themes* **2016,** *8*(4), 444–460. https://doi.org/10.1108/WHATT-04-2016-0022
The Economic Times, Slow Food: Indian Consumer Rediscovering a Taste for Local Ingredients, Over the Aspirational Imported [Online] 2015. https://economictimes.indiatimes.com/magazines/panache/slow-food-indian-consumer-rediscovering-a-taste-for-local-ingredients-over-the-aspirational-imported/articleshow/46890595.cms?from=mdr (accessed May 22, 2018).
The Hindu, Organized Retail in India to Double by 2021 [Online] 2019. https://www.thehindu.com/business/organised-retail-in-india-may-double-by-2021/article26379089.ece (accessed Feb 28, 2019).
The Tribune, India's Turn to Go Slow [Online] 2019. https://www.tribuneindia.com/news/spectrum/india-s-turn-to-go-slow/690444.html (accessed August 8, 2019).
Tasting India Symposium, Swaad [Online] 2018. http://www.tastingindiasymposium.com/wp-content/uploads/2019/01/SWAAD-International-Day-of-Indian-Gastronomy.pdf (accessed Dec 18, 2018).

CHAPTER 3

Understanding Indian Food Service Customers and Their Consumption Behavior

ABSTRACT

The chapter is all about understanding the Indian F&B (food and beverage) service customers and the consumption patterns and emerging trends. Given the regional, ethnic, socio-cultural, and economic diversity at hand, segmenting the customer base becomes a complex issue. The chapter deploys the research documented by the National Restaurant Association of India to do just that. The segmentation of the market and building an understanding of what drives the buying behavior is also discussed in the chapter. India-specific aspects around alcoholic beverage consumption and response toward the legal challenges too are taken up. The core of this chapter is focused on the typology of the Indian food and beverage customers and what drives them to buy F&B services. While the data cited and analyzed is specific to the organized formats of the massive Indian F&B service market, it is apparent that a lot of what is true for the organized formats are going to be relevant for the unorganized part of the business, especially as the trend is clearly headed toward organized segments.

3.1 INDIAN CUSTOMER: A DEMANDING NUT TO CRACK

True to its status as a developing economy, the Indian market represents an exciting opportunity for a range of product and service categories, and this definitely includes the F&B (food and beverage) service sector. The opportunity is defined by many aspects including the size of the potential market both in terms of the number of potential consumers as well as the cultural and socio-economic changes that are redefining how the huge population consumes F&B services. Liberalization of the Indian economy in the early 1990s set into motion policy-level decisions that have since impacted almost every facet of economic activity and associated consumerism growth.

As discussed in the earlier two chapters a range of segments have found roots, drawing from Porter's work on business strategy, it is quite apparent that the sector is no longer in infancy but rather in a growth phase. In fact, as the growth phase of key subsegments is analyzed further; the variables that are and will continue to impact the expansion become visible. Factors like demographics, location (region, metro–nonmetro, rural–urban, and tier 1/2/3), and other socio-economic and cultural elements are playing out and defining the customer response that the various subsegments of both (chain and standalone) formats of organized food services in India. What is quite clear is that the sector can look ahead to a long, though complex and demanding growth phase ahead. It will be a while before the maturity phase kicks in as far as the Indian market is concerned. Quite like any other industry in a growth spiral, the opportunity presents itself to the entry of competition, given that some subsegments require limited capital investment than others the barriers to entry become even more fragile. The ability to innovate, create value for current and future customers, and demonstrate dynamic capability in an ever-changing marketplace is the only way to build a sustainable competitive advantage; that is, if something like that even exists any longer. Bharadwaj et al. (1993) in their research on sustained competitive advantage in the service industry recognize that given the competitive forces at play, sustainable competitive advantage can turn out to be a moving target or even worse a mirage.

3.2 TYPOLOGY OF THE INDIAN CONSUMER

The potential ways of identifying, classifying, and further segmenting the F&B service customers can be multiple; the National Hotel and Restaurant Association in its 2016 report has deployed psychographic profiling of the consumer segments that are essentially based on the reason for consumption. Srividhya (2014) is categorical in stating that segmenting the customer group further is the way forward to understanding their eating out or dining behavior. The key psychographic segments are as shared below:

- Family bonding seeker
- Fun seeker
- Socializer
- Discerning urban

A brief profile of each of the earlier cited segments is shared in Table 3.1 for further clarification (NRAI Report, 2016), while it helps to classify the target market; it would be naive to not recognize that there are no hard lines that separate on group or purpose from another. The same customer can be easily found to express his/her dining behavior for one or the entire earlier listed segment types over a period of time. Equally possible is the range of purposes being expressed within a single group of diners at particular eating out occurrence. That said the classification in general; despite the earlier stated complexity is an important aspect for F&B service operators to consider as they determine a fit between what they are offering and their target market segment(s).

TABLE 3.1 Psychographic Consumer Segments—Brief Profile.

Family bonding seeker	Family bonding seekers view eating out as an opportunity to get together with the family, an opportunity to cherish relationships. Family becomes the pivot around which all fun and celebration is centered. However, this group is relatively more conservative in their approach. They are risk-averse and usually look for places to eat that would work for everyone.
Fun seeker	Fun seekers are generally innovative and experimental and eat out in a group. Having fun is the order of the day and they seek new experience in everything they do. They wish to try new restaurants and new dining concepts.
Socializer	Socializers have a large network of acquaintances and view eating out as a form of entertainment and social interaction. Their emphasis is more on social relationships and they try to get whole group together.
Discerning urban	Discerning urban belong to higher-income households and look at eating out as a place to chill and unwind. They seek places to relax and are willing to pay a premium for quality. They tend to choose places to eat that define their social-economic status as much as the dining experience they offer.

Source: Adapted from NRAI Report (2016).

3.2.1 THE MARKET, THE CUSTOMER, AND THE REASONS TO BUY

The F&B services market is thriving, growing, and evolving constantly; the sheer size of the market given the population of the country being 1.36 billion people by end of 2018 gives one an idea of the potential, given the typical daily meals being breakfast, lunch, and dinner the simplistic expression of the two variables in the form of a number of daily meals being consumed comes to a staggering 4 billion plus, not accounting for the snack/beverage consumption aspect. The earlier observation is limited in many ways and fails to account for many variables that would need to be considered to get to a more realistic as well as actionable data for the industry. What it does do is give a feel for the potential "EOO" (eating out occasions) that the market can look at as a business opportunity for today and the future. The Times of India (2016) reported that urban Indians eat out at least six times in a month, and this statistic is only growing further. This is where it becomes important that the market segments are better understood, identified, and targeted by the industry in an informed manner. As stated earlier a society as diverse as that in India, where religion, caste, economic status, gender, and regional differences continue to define consumption behavior, the segmentation of the market itself becomes a tall order. To make this even more of a daunting task than it already is, is the pace of change that the society is undergoing, all of which is impacting and redefining the consumption behavior, making it hard for researchers and industry observers to hold their understanding of how a segment is behaving today with how that may change over a rather short duration of time in the future.

One can eat at home, work, and outside of both home- and work-based situations from a location of the "eating act" perspective. If one does "brown bagging" (taking food from home), which is quite a common practice in India traditionally, that represents a lost "EOO" for the industry in principle. So the more the potential customers cook and eat at home or carry home-cooked meals for consumption the more the potential revenue dilution for the F&B industry. This is what makes research into understanding eating out behavior a critical element for the growth strategy of the industry players. The conventional wisdom of working with demographic data that provides inputs like age, gender, geographical location, social status, income category/discretionary income, religion, and other such information; though helpful has limitations when it comes to segmenting the food services market and understanding eating out behavior of consumers. The following mentioned questions can be considered as the foundation of any research into the eating out/ordering in (delivery) behavior:

- Who is the customer? Which segment?
- With whom is the EOO/order in taking place?
- Why—what is the occasion?
- Where—which type of F&B service outlet is being patronized?
- What cuisine or food choice is it for?

Answers to the earlier stated questions can provide a holistic view of the consumer buying behavior; the National Hotel and Restaurant Association of India as one of the key bodies providing guidance to the food service industry in terms of research into buying behavior of the Indian customer recognizes the need for the investors as well as operators to remain focused on the five variables to remain clued to shifts in consumer behavior in a fast-changing and ever-evolving dynamic market.

Being able to understand your target customer and what he/she values is a good start. Sustainable competitive advantage in the business however will involve both adding new customers as well as retaining the old ones. This will require a food service business to ensure that the value proposition is retained such that customers remain brand loyal. By all means, a tough ask given the rate transparency and the digital ease with which competition can reach the same customers. Bowden et al. (2013) articulate this challenge by looking at satisfaction, trust, and customer delight as the key drivers of loyal buying behavior; this factors in service quality implications, ability to generate "WOW" moments representing customer delight, and eventually trust that the "brand" always delivers the promise. Success in food service business will need a holistic approach based on a sound understanding of the customer expectations and by extension their buying behavior.

3.2.2 THE INDIAN CONSUMER'S EATING OUT OPTIONS AT A GLANCE

Each of the following mentioned F&B service segments has a purpose to serve, a value to offer to a range of customers and for a range of reasons. It is imperative for current operators as well as potential investors (domestic as well as international) to understand how the consumers are patronizing the wide range of F&B service formats that are now available to them and are vying for the revenue rupee aggressively, trying to attract each "eat out occasion possible," and encouraging them to eat out more and more as well (Table 3.2) (NRAI Report, 2016).

TABLE 3.2 Format-wise Eating Out Preferences of Indian Consumer.

Format	% EOO share	Total EOO in Mn
Cafe	17%	600
QSR	36%	1260
CDR	30%	1050
FDR	1%	35
PBCL	5%	175
FD/IC	11%	380

Source: Adapted from NRAI Report (2016).

The ***cafes*** emerged in the metropolitan cities and quickly moved into a wide range of tier 1 and 2 cities and are expanding to more cities. The café chain market alone will be worth INR 4500 crore by 2023 given the pace of growth (Economic Times.com, 2019). From meeting friends and families to just hanging out for reading or even spending time alone, the cafes offer the perfect spaces with a range of brands offering different levels of services and ambiance. Even within the segment, there are subsegments; upmarket cafes are good places for business meetings. The national favorite (as a reason to visit) is to organize interaction between families for the purpose of an arranged marriage. For the middle-class Indian, a café offers a price-sensitive neutral turf for the boy–girl meeting with parents often sitting separately to evaluate any potential for the arranged marriage proposal to go forward. While branded chains offer superior ambiance and standardized products, making them amenable to growth via the chain format; nonbranded options are available in plenty as well. That said the Indian market is clearly a hot spot for international investment in this space (Restaurant India.in, 2018). While key brands and their growth have been discussed in earlier chapters, the sheer cultural significance of this segment is worth discussing as that is what drives consumption behavior and hence must be understood.

The ***QSRs (quick service restaurants)*** as per the data shared earlier account for 36% of the EOO (1260 million) based on research conducted by NRAI (National Restaurant Association of India). The segment lives to its name by offering quick service, prices that the large Indian middle class can afford, a range of F&B categories that work for the diverse needs of equally diverse customers. The demographic shifts in the Indian society have a lot to do with the emergence and growth of QSRs, increasing number of nuclear families with both spouses working, expanding spending power, time constraints socio-cultural dynamics have resulted in a shift from cook at home, eat at home to a "grab a bite" on the go trend, one that is now catching up even in the smaller cities and towns. From Birthday celebrations to a casual dine out with family, friends, etc., the QSRs have built patronage from a range of customer types for a range of reasons. As international brands like Macdonald's, Pizza Hut made their way to the Indian market. It was initially seen as aspirational to try their offerings, almost like the nation was waiting for them to arrive, even though that phase is gone and exposure to international brands is there now, a generation of young Indians has been hooked to the food and the environment in which it is served. The investment just keeps pouring in as far as the QSR space is concerned (Food and Hospitality.in, 2017).

CDRs (casual dining restaurants) with a 30% share of the EOO spread (1050 million) have to ensure that they are able to segregate their offerings for the customers as they compete with the QSR space, given that the QSRs are likely to be more self-service oriented; the CDR segment usually offers varying levels of service such that the price point (average check per customer) remains attractive to a larger customer base, the food choices guided by the typical "multi-cuisine" approach that offers something for every potential customer has worked for the segment. As discussed in earlier chapters, the CDR segment of today has evolved from the fragmented unorganized "restaurant" of yesteryears; the segment has responded to the needs of the customer today and is now offering services to families for outings and special event celebrations, venue for business meetings and team get-togethers, and so on. Care Ratings report on Indian Restaurant and Food Service Industry (2019) recognizes that the QSR and CDR formats will drive the market penetration of the organized food and service formats over the next decade.

Between them the QSR and CDR segments account for more than 65% EOO market share, thus cementing their significance in that respect.

PBCL (pubs, bar cafes, and lounges) though more of a big city phenomenon represents a huge socio-cultural shift for India, with 175 million EOO at 5% of the total volume (NRAI Report, 2016). These outlets are attracting higher-end spending professionals as places to relax after work, for informal meetings, and even as spaces for celebrations. With a largely young customer base, this is one segment that is redefining the socio-cultural norms that were quite well entrenched till a few years ago. Another dimension worth mentioning here is the changing gender equations that the segment seems to be leveraging and fuelling simultaneously. The female Indian professionals of today are not chained to past moralities and social norms that defined their F&B consumption behavior; they are entertaining and being entertained by friends, colleagues, family, and can be seen openly consuming alcoholic beverages with gusto. IMARC report (2018) sees the combined value of the cafe, pub, and bar formats reach 3.5 billion USD by 2024.

There are of course sharp lines of divide in the Indian society with respect to consumption of alcohol; let alone having women consume in public places, thus the earlier discussed pattern of consumption behavior is still confined to just the metro cities. Political, religious, and other complex socio-cultural systems with multiple layers are at work to question what is often described and portrayed as west influenced the corruption of Indian values and morality. Prohibition in states like Bihar, Tripura, and longstanding dry-status of Gujarat is as much a reality as is the growth of pubs and bar cafes. However, one can safely argue that social drinking among women has found roots in India and is here to stay as a trend, at least for the social elite and professionals working in the top tier cities (Business World.in, 2019).

It is no surprise that the **FDRs (fine dining restaurants)** account for as less as 1% of the EOO, one must however not be misled by the number as the revenue impact of the segment cannot be ignored. Another aspect that is relevant is the role this segment plays in defining the service as well as a food-related benchmark that the industry stands for. In comparison to the Western European and North American market where FDRs are rated, recognized, and celebrated through awards; the Indian landscape of FDRs is still rather limited. Though a few restaurants have received national, if not international acclaim; the number of FDRs is not comparable and the quality of the offerings is a work in progress. While the segment has come a long way from a period in time when the FDRs could mostly be found in five-star hotels based in the metros to quite a few standalone fine ding restaurants finding presence in the metro and tier 1 cities across the nation. The Indian customer does have the purchasing power as well as the desire to experience the best of service and food product that restaurant can offer, for business meetings to family celebrations to time out with friends; the social elite are more than willing to express their financial and social stature by frequenting FDRs.

FD and IC (frozen desserts and ice cream) bars too have grown in recent years into popular EOO spots; brands like Keventers and many other international, as well as national brands, are in play to capture the generation Z as the country moves from the traditional formats to more expensive but service-oriented ones.

3.3 WHY DO INDIANS EAT OUT AND WITH WHOM DO THEY DO THAT

Reasons that get Indians to eat out and patronize the earlier discussed F&B segments are important to identify and understand. The research conducted by NRAI and documented in their 2016 report provides data on not just the typical reasons but also the spread in percentage terms to provide the industry users a better perspective on how the 10 reasons mentioned in Table 3.3 pan out (NRAI Report, 2016).

A review of the earlier shared data provides perspective on a rather interesting spread (not that this is going to be much different for any other part of the world); the reason "alternate to cooking" is representative of a socio-cultural shift that is still limited to a very minuscule part of the population, as the joint family system which was quite the norm in India continues to get diluted, as more and more nuclear, double-income families find presence, this reason to eat out is just warming up. One can only imagine the consumption dynamics if this trend expands and grows to the tier 2 and 3 cities and semi-urban India.

The two other interesting reasons that have emerged are that of the "experimental eater" and the "compulsive eater" at a combined 16% weight, these are interesting consumer groups to look at; in fact, it would be interesting if more research focused on these two segments is done to get clarity on their F&B service segment preference as well as their capacity to spend. This would be actionable input for the industry.

TABLE 3.3 Reasons for Eating Out: Percentage Split.

11%	14%	14%	8%	15%	6%	7%	8%	8%	8%
Family functions	Family outing	Friends outing	Business meeting	Shopping and/or casual outing	Festival celebrations	Spending free time	Alternate to cooking	Compulsive eating out	Experimentation

Source: Adapted from NRAI Report (2016).

Family and friends related reasons, as well as festival-related celebrations are generic reasons to eat out across the world; no surprise that the NRAI, 2016 report data on "who do you eat out with" gives a massive 56% reference point to family and another 31% to friends as the key social groups that Indians find reason to go out and eat with. Table 3.4 shared below also documents the strong consumer leaning toward the QSR and CDR formats as the chosen F&B service options while eating out (NRAI Report, 2016).

The preferred choice of eating out behavior has another layer to be considered; that is, the timing—when in terms of days (weekdays/ weekend days) and what part of the day for what type of social group:

- Friday evenings, weekends are for family-related eat out occasions.
- Colleagues eat out mostly in the afternoons during weekdays, but evenings are getting increased play.
- Friends find time without clear time limitations.

TABLE 3.4 F&B Service Format Choice of Accompanying Group.

Social group	% of EOO share	Choice of format
Family	56%	CDR
Friends	31%	QSR
Colleague	9%	CDR
Alone	4%	QSR

Source: Adapted from NRAI Report (2016).

3.3.1 BEVERAGE CONSUMPTION BEHAVIOR IS THE TREND TO LOOK OUT FOR

It is not lost on the industry operators that the "beverage" component of the product spread that various formats and subsegments have on offer are a critical and often more profitable part of their menu mix. The significance of how beverage consumption behavior is shifting in the Indian market is tremendous, segments like cafes and the PBCL are beverage centric to start with, and understanding consumer preferences in terms of where they stand today and where they are headed as a trend is always going to helpful to operators as they innovate and try and offer products that are well received by the customer base.

It was not too far back that beverages were offered in restaurants as a tool to up-sell and thus generate incremental revenue that would help push the average check with an obvious intent to maximize revenue. The service staff across restaurant operations (CDRs and FDRs) was briefed by managers to recognize the need to push beverages (alcoholic or otherwise) to each customer as an essential part of the order taking routine. Soon enough the industry segments got familiar with the opportunity the price bundling presented and started packaging food–beverage combos at a price that created value and encouraged the customer to spend a little more to get a holistic meal experience with appropriate beverage options in tow. The QSR format too did the same; in fact, even more successfully by offering a range of meal combinations not just for individuals but even family packs, children packs, etc.

Earlier conversation in this book on the "taboo" around the consumption of alcoholic beverages in India; the religious and gender-related factors, and the politics around it, all are relevant here as well. As Indian society continues to open up with respect to the acceptance of social drinking (alcoholic beverages); women exercise the right to make personal consumption choices, their impact on "beverage" as a product category will continue to change as well. That said each F&B service segment will also have its unique experience of ordering behavior for nonalcoholic and alcoholic beverages. Table 3.5 provides guidance with respect to the nonalcoholic beverages component (NRAI Report, 2016).

TABLE 3.5 Nonalcoholic Beverage Consumption Choices across F&B Service Formats.

Format	Preferred beverage (top three and their % share)		
Café	Coffee 30%	Tea 20%	Cold coffee/iced tea 20%
QSR	Aerated beverages 46%	Shakes and smoothies 10%	Branded nonaerated drinks 10%
CDR	Aerated beverages 30%	Mocktails, ethnic drinks, and Others 28%	Juice 15%
FDR	Others 29%	Juice 20%	Aerated beverages 15%

Source: Adapted from NRAI Report (2016).

Research also suggests that nearly a quarter of customers order an alcoholic beverage when eating out with varying frequency. This trend is largely visible in the metros and mini metro cities, but as the socio-cultural dynamics continue to shift this trend will continue. Another interesting phenomenon around the consumption of alcoholic beverages is the **"BYOB"** (bring your own booze) system; this approach offers enablement for alcohol consumption by patrons at F&B service locations at two levels, the first is the price advantage; the customer can bring in his/her own alcoholic beverage (most BYOB joints have a liquor shop right outside) and then just buy food that is served at the outlet and any nonalcoholic beverages or mixers that may be required. Some locations charge a service charge or additional charges for glassware, use and throw cups/glasses, etc. The concept of charging "corkage" is almost nonexistent, except for some hotels or caterers using the same to deter customers from bringing their own booze rather than getting the service from the event operator/hotel. The prohibitive cost of getting liquor license has got everything to do with the BYOB concept and its popularity. This system helps bypass the legal system and also brings down the landed cost to the customer; in many ways, the F&B joints operating along this framework undercut the high-end bars, pubs, and restaurants. There is a clear segment separation herein; the price-sensitive customers (usually younger professionals) go the BYOB way, while the more service conscious, high spending ones go the other way. It is quite an Indian way to resolve what law hinders by finding a grey area that works for all stakeholders. The liquor shops around the BYOB restaurants too are seldom licensed; most however enjoy political patronage and hence do brisk business law notwithstanding (Figs. 3.1 and 3.2) offer a visual representation of the BYOB concept.

3.3.2 ALCOHOL CONSUMPTION AGE DILEMMA IN INDIA

Another interesting facet of alcohol consumption in India is the legal age issue. Different states have defined different ages as allowable for the consumption of alcohol; in Goa, it is 18 years; while some like Assam and Andhra Pradesh have it at 21; Delhi and Maharashtra have settled for 25; and Gujarat, Tripura, and Bihar have made it outright illegal to consume alcohol. It is hard to wrap ones head around the logic for this variation; there may have been some rationale to start with as the nation became independent, but it does look like almost similar to the need for tax law rationalization this issue too needs a close look.

FIGURE 3.1 BYOB concept at play in India.

Source: Author.

FIGURE 3.2 Typical interior setup of a BYOB bar/restaurant.

Source: Author.

The sheer impunity with which the age restrictions are ignored by consumers and service providers should inspire the powers that be to consider an amendment to an archaic law or figure a way out to give teeth toward its enforcement. The recent litigation in 2017–2018 that led to the ridiculous regulation, that presented retail/service of alcohol within 500 meters of a national highway almost got the industry to its knees overnight as businesses tried in vain to find solutions. Some did find innovative solutions, while others could not; either way courts demonstrated how legal overreach can cause damage to the industry without any real benefit or fulfillment of any purpose.

In the shadow of the challenges unique to India and situations discussed earlier, alcoholic beverage consumption is charting steady growth, with the gender barriers coming down; social-cultural change gaining ground this growth will stay the course. While wines are popular among women, pubs are coming up with fresh beer to the delight of a growing group of beer enthusiasts. That said from Vodka, Scotch, and other spirits to Tequila, the game is on. Different sources will provide different statistics on which alcohol category is ruling the roost, but it is apparent that seasonality is a factor and the trends are as dynamic as the market; the generation Z is warming up to social drinking and that is the one trend that the industry is capitalizing on. Innovative pricing approaches in the form of "bar exchanges" that builds in revenue management practices that were till yesterday confined to hotel accommodation or airline bookings are being well received by the F&B segment. Another chapter will look at the practice of revenue management in the F&B service space as also the digital marketing footprint of the sector.

3.3.3 ORDERING F&B: HOME DELIVERY OR TAKE AWAY

The sheer ease with which consumers can today use their cell phones to search for F&B service options, use various apps to select items from the menus of their favorite food joints, place orders, and make payments in a seamless manner has given an impetus to the pickup-delivery models. It is common to see F&B service business across nearly all segments to make investments to enable a home delivery service; this service is usually free of cost. However, it is not uncommon for businesses to have riders like "minimum purchase value" or "maximum delivery distance" for which no extra charges are levied and then charge customers a delivery charge where the riders are not met. Some QSR brands also offer delivery time guarantee (Domino's, Pizza Hut); wherein, if the service standard is not met then the customer is compensated along predetermined lines (up to Rs. 500 waived etc.).

NRAI (2016) report documents that of all the delivery orders made, nearly 80% are made telephonically while the rest are made online either through business websites, food aggregator website, or mobile apps. The report also mentions that factors affecting the choice of service provider include the brand visibility, time taken for delivery, and finally any discounts being offered. Apps like "Gourmet Passport," "Zomato Gold" among others offer discounts at a wide range of F&B operations. The research also cites data that establishes an ordering frequency of 1.5 orders per month per household for the metros, 1.3 for mini metros, and 0.8 for the tier 1 and 2 cities in India. It was also mentioned that weekends found more ordering activity as compared to weekdays, and also late afternoons–evening time was the peak time for businesses in terms of food order deliveries.

The logistics back end behind the delivery models for F&B products home delivery as well as take away options have come a long way in terms of deployment of technology for guidance in terms of directions, sending electronic status updates to customers, packaging that facilitates temperature control as well as convenience for the customer is quite in focus; though dependent on the segment being serviced with the price point guiding the quality of packaging being deployed. The manpower required to deliver is not really an issue for a nation with ample young unemployed looking for employment opportunities. Most of the businesses hire delivery boys, who have their own two-wheelers that they can use for deliveries; they often pay them wages plus cover their mileage costs. Some brands build their own fleet of two-wheelers to manage the delivery back end. These vehicles are also used for brand advertisement purposes, and it is common for the delivery boys to wear uniforms that represent the brands they are working for.

Drive through option is not that well-entrenched as a delivery model but a few international fast-food brands (Macdonald's) have tried the option in metro cities, not quite the preferred option as far as the Indian market is concerned. This may have more to do with the road infrastructure as well as general driving conditions than the option. It is much easier to just order a home delivery.

3.4 CONSUMPTION PATTERNS OF THE KEY PSYCHOGRAPHIC SEGMENTS

3.4.1 FAMILY BONDING SEEKER

Family bonding seeker represents potentially the largest consumer segment. Culturally India continues to be a family-centric society, even though larger and more complex joint family systems have seen erosion of sorts and the smaller nuclear families have grown, eating together, at home or otherwise remains an important cultural construct. It is no surprise that eat out (or calling in food at home) for the purpose of family bonding has become a major segment drawing the attention of the F&B brands (Tables 3.6 and 3.7) (NRAI Report, 2016).

> For the family bonding seeker, the informal setting of the QSR and CDR formats seems to work well as these two segments account for nearly 70% of the EOO. Also, no surprise that this psychographic segment enjoys nearly 90% of the EOO with family and friends. While there are many reasons to eat out, family get-togethers, special events like birthdays, anniversaries top the list.
>
> The family bonding seekers tend to spend about INR 5000–6000 per month, though the frequency of eating out is about 2–3 times a month given the planned nature of the occasions and the size of the group, the overall spend is quite high. Choice of cuisines seems to be restricted to Indian and Chinese, but the trend in transition.

TABLE 3.6 F&B Format Preferences of Family Bonding Seeker.

Format	% EOO share
Café	11%
QSR	37%
CDR	31%
FDR	1%
PBCL	3%
FD/IC	17%

Source: Adapted from NRAI Report (2016).

TABLE 3.7 Key Factors for Brands/Outlet Selection for a Family Bonding Seeker.

Most important	Important	Least important
Taste/quality	Location/approachability	Menu/variety
Ambience and hygiene	Quality of service	Promotional offers
Familiarity with restaurant	Price affordability	
	Parking facility	

Source: Adapted from NRAI Report (2016).

3.4.2 FUN SEEKER

This psychographic segment is really about a group of consumers that is open to experimentation; they are foodies and by definition always looking try new places to eat, new types of food and this approach constitutes the "fun" part of their lives. The changing socio-economic and cultural landscape of India has everything to do with this customer profile. A younger group, with purchasing power, disposable income that is growing, and interest in socialization over F&B is what defines the "fun seeker" segment (Tables 3.8–3.10) (NRAI Report, 2016).

For the fun seeker, the informal setting of the café, QSR, and CDR formats seems to work well as these three segments account for nearly 85% of the EOO. Also, no surprise that this psychographic segment too enjoys nearly 85% of the EOO with family and friends. Eating with colleagues and even alone is quite acceptable for this group.

The fun seekers tend to spend about INR 5000–6000 per month per household, though the frequency of eating out is about 5–6 times a month. This group is more price-conscious; choice of cuisines seems to be wider with fast food, bakery, confectionary are popular, but the segment is open to Indian and Chinese options too.

TABLE 3.8 F&B Format Preferences for Fun Seeker.

Format	% EOO share
Café	17%
QSR	45%
CDR	21%
FDR	<1%
PBCL	5%
FD/IC	12%

Source: Adapted from NRAI Report (2016).

TABLE 3.9 Accompanying Group Preferences for Fun Seeker.

Accompanying	% EOO share	Preferred format
Family	53%	CDR
Friends	32%	QSR
Colleague	8%	Café
Alone	7%	QSR

Source: Adapted from NRAI Report (2016).

TABLE 3.10 Key Factors for Brands/Outlet Selection for a Fun Seeker.

Most important	Important	Least important
Taste/quality	Promotional offers	Familiarity with restaurant
Manu/variety	Ambience and hygiene	Location/approachability
Price affordability	Quality of service	Parking facility

Source: Adapted from NRAI Report (2016).

3.4.3 SOCIALIZER

This psychographic segment is focused on building and sustaining a social network, a part of this orientation is linked with entertainment over F&B (Tables 3.11–3.13) (NRAI Report, 2016).

TABLE 3.11 F&B Format Preferences for Socializer.

Format	% EOO share
Café	15%
QSR	41%
CDR	36%
FDR	1%
PBCL	4%
FD/IC	3%

Source: Adapted from NRAI Report (2016).

TABLE 3.12 Accompanying Group Preferences for Socializer.

Accompanying group	% EOO share	Preferred format
Family	53%	QSR
Friends	32%	CDR
Colleague	13%	CDR
Alone	2%	QSR

Source: Adapted from NRAI Report (2016).

TABLE 3.13 Key factors for Brands/Outlet Selection for a Socializer.

Most important	Important	Least important
Quality of service	Familiarity with restaurant	Promotional offers
Ambience and hygiene	Price affordability	Location/approachability
Taste/quality	Menu/variety	
	Parking facility	

Source: Adapted from NRAI Report (2016).

3.4.4 DISCERNING URBAN

This psychographic segment has been in existence since pre-independence India representing the minuscule part of the population with the financial means to demand the very best in hospitality with respect to both food as well as the accommodation elements. They constitute the elite of society. They express their "taste" through the consumption preferences that they exercise. What is interesting about this segment of consumers now as compared to earlier is the sheer size of the segment in terms of number of consumers who may classify as consumers willing to pay a premium price for the superior quality of products and services. This is a direct result of the economic expansion witnessed by the country and also the exposure to global F&B trends (Tables 3.14–3.16) (NRAI Report, 2016).

For the discerning urban in addition to the café, QSR, and CDR formats, the FDR and PBCL formats also find play, also they tend to choose outlets and brands offering superior quality. This psychographic segment too enjoys nearly 85% of the EOO with family and friends. Eating with colleagues also works, but eating alone is rather limited.

The discerning urban tend to spend about INR 8000–10,000 per month per household, though the frequency of eating out is about 3–4 times a month. This group is focused on the quality of experience, but eating out is an important social activity to express their status; for them choice of cuisines seems to be Italian, pan-Asian, Continental and Indian, and some regional food being welcomed as well.

TABLE 3.14 F&B Service Format Preferences for Discerning Urban.

Café	24%
QSR	20%
CDR	37%
FDR	3%
PBCL	8%
FD/IC	8%

Source: Adapted from NRAI Report (2016).

TABLE 3.15 Group Preferences for Discerning Urban.

Accompanying group	% EOO share	Preferred format
Family	47%	CDR
Friends	33%	CDR
Colleague	14%	CDR
Alone	6%	CDR

Source: Adapted from NRAI Report (2016).

TABLE 3.16 Key Factors for Brands/Outlet Selection for a Discerning Urban.

Most important	Important	Least important
Taste/quality	Menu/variety	Price affordability
Quality of service	Familiarity with restaurant	Promotional offers
Ambience and hygiene	Location/approachability	
	Parking facility	

Source: Adapted from NRAI Report (2016).

3.5 CONCLUSION

At the end of the day it boils down to getting a good sense of how the various consumer segments and subsegments are expressing their F&B purchase requirements and choices. The chapter has utilized available research shared by the NRAI to offer a perspective that tries to critically examine the consumption behavior by answering the following set of questions:

- Who is the customer? Which segment?
- With whom is the EOO/order in taking place?
- Why—what is the occasion?
- Where—which type of F&B service outlet is being patronized?
- What cuisine or food choice is it for?

The summary shared in Tables 3.17 and 3.18 establishes that as far as the organized F&B service segments are concerned, the QSRs, CDRs, and the cafés account for more than 80% of the EOO. This spread remains true for metro, mini metro and tier 1 and 2 cities in the country; the only clear shift seems to be the dominance of the CDRs for the tier 1 and 2 cities. The Mint (2016) in a post in this respect states that while Indians today do not need a reason to eat out; when they do, they still look for casual, value for money and food that is closer to home food.

TABLE 3.17 F&B Service Format preference by City Type.

Where?	India	Metro	Mini metro	Tier 1 and 2
Café	16%	16%	16%	13%
QSR	37%	37%	48%	31%
CDR	30%	25%	21%	40%
FDR	1%	1%	1%	<1%
PBCL	5%	7%	4%	4%
FD /IC	11%	14%	10%	11%

Source: Adapted from NRAI Report (2016).

TABLE 3.18 F&B Service Format Preferences—(% EOO Share).

With whom?	India	Metro	Mini metro	Tier 1 and 2
Family	56%	52%	53%	58%
Friends	31%	35%	33%	30%
Colleague	9%	10%	8%	9%
Alone	4%	3%	6%	3%

Source: Adapted from NRAI Report (2016).

More than 85% of the EOO involve eating with friends and family, though nearly 10% of the EOO are finding colleagues eating together; this trend seems to be growing. It is also apparent that the socially elite are looking at cuisine preferences beyond the local Indian/regional options and are enjoying continental, Italian, pan-Asian food choices, in addition to the typically Indian food. The branded fast food (pizza, burgers, etc.) is well entrenched across all consumer segments; the only difference that socio-economic classification may be causing the choice of brand, with price-sensitive consumers enjoying the local brands and the rest patronizing the international brands, many of which have customized their offerings to meet the needs of the Indian customers over the years (Tables 3.17 and 3.18).

As discussed in the earlier chapters, the organized F&B service sector in India has grown significantly over the last decade or so; that said the unorganized format still remains relevant and massive, especially with respect to the rural and semiurban parts of the country. The psychographic segmentation of the Indian customer as discussed here also finds application for a larger understanding of the Indian consumer even though the data analyzed and discussed represents only the organized formats of the business. Given the market size, the diversity that India represents in terms of culture, religious aspects of social expression, economic segregation that represents extremes, understanding what drives customer consumption behavior will remain important for both current operators in the F&B services space as well as potential inventors who may be looking to invest in this growth story.

A lot of the discussion in the chapter is about the response of the Indian consumers toward the wide range of organized F&B service formats that are vying for their purchase, the organized format currently represents less than 40% of the Indian F&B service market, the remaining is the huge unorganized segment. Do customers in the Indian rural areas and semiurban towns have the same reasons to patronize F&B outlets (unorganized) or can we expect differences? While some parallels can be argued easily, substantial differences are to be expected. Some of these aspects would be a result of socio-cultural-economic setting, but equally relevant would be the availability of options in these parts of the country. As the growth curve of the organized formats extends there penetration to these parts of the country, it would be interesting to observe how both the operators and customers respond to each other.

KEYWORDS

- **psychographic F&B segments**
- **fun seeker**
- **family bond seeker**
- **F&B service segments**
- **Indian F&B customer**

REFERENCES

Asp, E. H. Factors Affecting Food Decisions Made by Individual Consumers. *Food Policy* **1999,** *24* (2–3), 287–294.

Auty, S. Consumer Choice and Segmentation in the Restaurant Industry. *Serv. Ind. J.* **1992,** *12* (3), 324–339.

Bharadwaj, S. G.; Varadarajan, P. R.; Fahy, J. Sustainable Competitive Advantage in Service Industries: A Conceptual Model and Research Propositions. *J. Mark.* **1993,** *57* (4), 83–99.

Bowden, J. L.; Dagger, T. S. To Delight or Not to Delight? An Investigation of Loyalty Formation in the Restaurant Industry. *J. Hosp. Mark. Manage.* **2011,** *20* (5), 501–524.

Bowden-Everson, J. L. H.; Dagger, T. S.; Elliott, G. Engaging Customers for Loyalty in the Restaurant Industry: The Role of Satisfaction, Trust, and Delight. *J. Foodserv. Bus. Res.* **2013,** *16* (1), 52–75.

Engel, J. F.; Blackwell, R. D. *Consumer Behavior*; Dryden Presee: New York, NY, 1982.

Euromonitor Report (2019) on Cafes and Bars in India.

Express Food and Hospitality, QSR Market Investment Growth in India, [Online] 2017, Available from: https://www.foodhospitality.in/food-retail/qsr-market-investment-growth-in-india/385704/, (accessed Aug 12, 2018).

Fishbein, M.; Ajzen, I. *Belief, Attitude, Intention and Behavior: An Introduction to Theory and Research*; Addison-Wesley: Reading, MA, 1975.

FICCI Report. Unlocking the Potential in Food and Beverage Services Sector. [Online] 2015, http://ficci.in/spdocument/20589/F&B-Report.pdf (accessed June 12, 2018).

Food Service India. Key Trends Impacting Food Service Industry in India. [Online] 2019, Available from: https://www.indiaretailing.com/2019/01/17/food/food-service/expert-speak-key-trends-impacting-foodservice-industry-in-india/ (accessed May 10, 2019).

Hospitality Trend Report. Published by AF and Company Consulting IBEF.ORG, India Brand Equity Foundation. [Online] 2018, https://www.ibef.org/download/Tourism-and-Hospitality-January-2018.pdf (accessed Apr 20, 2018).

IBEF.ORG, India Brand Equity Foundation. [Online] 2017, https://www.ibef.org/industry/retail-india.aspx (accessed Jan 4, 2019).

Indian Restaurant & Food Service Industry–Structure & Prospects. [Online] 2019. Available from: http://www.careratings.com/upload/NewsFiles/Studies/Restaurants%20%20QSRs%20May%202019.pdf (accessed Jul 10, 2019).

IMARC Group.com. India Pub, Bar, Café and Lounge (PBCL) Market: Industry Trends, Share, Size, Growth, Opportunity and Forecast 2019–2024. [Online] 2018. Available from: https://www.imarcgroup.com/india-pub-bar-cafe-lounge-market, (accessed Jan 10 2019).

India Hospitality Industry Review. [Online] 2018, Published by HVS- Anarock, https://www.hvs.com/article/8488-hvs-anarock-india-hospitality-industry-review-2018 (accessed Apr 8, 2019).

Johns, N.; Pine, R. Consumer Behavior in the Food Service Industry: A Review. *Int. J. Hosp. Manage.* **2002,** *21* (2), 119–134.

JLL, India Hospitality Review. [Online] 2017, http://www.jll.co.in/india/en-gb/Research/Ready-steady-go-India-hospitality-industry-review-2017.pdf (accessed May 10, 2018).

Munjal, S.; Bhushan, S. *The Indian Hospitality Industry: Dynamics and Future Trends*, Apple Academic Press: New Jersey, USA, 2017.

NRAI. India Food Services Report, Published by National Restaurant Association of India, 2016.

Prasad, C. J.; Aryasri, A. R. Effect of Shopper Attributes on Retail Format Choice Behavior for Food and Grocery Retailing in India. *Int. J. Retail Distrib. Manage.* **2011,** *39* (1), 68–86.

Srividhya, N. Eating Out Habit of Individuals – An Analytical Study, *Int. J. Bus. Manage. Invention* **2014,** *3* (6), 38–44.

The Economics Times.com. Indian Café Chain Market will Reach Rs 4,540 cr by 2023. [Online] 2019. Available from: https://economictimes.indiatimes.com/industry/services/hotels-/-restaurants/indian-caf-chain-market-will-reach-rs-4540-cr-by-2023-report/articleshow/69997833.cms?from=mdr (accessed Jul 5, 2019).

The Times of India. Indians Eat Out at Least 6 Times a Month. [Online] 2019, Available from: https://timesofindia.indiatimes.com/business/india-business/indians-eat-out-at-least-6-times-in-a-month/articleshow/69318801.cms (accessed June 12, 2019).

The Live Mint. [Online] 2016, Available from: https://www.livemint.com/Consumer/DeU2aNxRI3U0unsxnjhy5L/Study-Quickservice-restaurants-top-dining-destination-for.html, (accessed June 7, 2017).

CHAPTER 4

Challenges, Issues, and Opportunities in the Food and Beverage Services Business in India

ABSTRACT

The chapter looks at the key issues and challenges that the food and beverage service sector in India grappling with. From a perspective of operators and potential investors alike, having a clear understanding of some of the deep-rooted problems that need to tackled while operating businesses in India is a must. There are aspects which are systemic, like the strong presence of the unorganized F&B service establishments that will take time to get resolved, there are others that are policy linked, which is where governmental action can make a quick difference. The chapter discusses both types. The author also identifies some emerging trends in the market they may be currently finding traction in the urban parts of the country but their sway is expected to expand. These are essentially opportunities that are opening up under the backdrop of the issues that cause the industry to find innovative solutions to achieve growth and sustainability.

4.1 INTRODUCTION

As would have become apparent to any reader the food and beverage services businesses in India have unique challenges and issues to grapple with, what makes it worthwhile to invest this domain is the plethora of opportunities that this high growth space presents. Although some issues and challenges are more of legacy elements that the industry seems to be carrying for reasons that are quite debatable at best, the changing trends and market dynamics bring more challenges and issues on the road as operators and investors compete in a sometimes quite crowded space. For the longest time, food and beverage service space has witnessed a success rate of about 30% when it comes to the number of new entrants found standing after a year, 3 out of 10 make it and only one is generating above-average profits, this is a rather loose assertion to make, especially without data to back it up and also devoid of any implication of variability in outcome based on what type or segment of food and beverage service being discussed. A wide spectrum of experienced professionals and operators are likely to nod at this assertion and further comment that the business is complex, competitive and there are just too many variables that need to sync together for it to work. Table 4.1 provides a snapshot of financials for four major listed players in the Indian food and beverage service market, though by no means conclusive, the data does provide some guidance on where the industry is headed (Care Ratings Report, 2019). The operating margins seem to have recovered from the shock of demonetization and goods and services tax and are not at 17%, but the net profit is still quite sluggish at less than 5%, finance charges (interest cost) seems to be diluting the profitability of the listed businesses in a significant way.

TABLE 4.1 Financial Highlights for Listed Restaurant Companies.

Key financial indicators	2017	2018	2019
Operating profit margin (%)	13.8	15.3	16.7
Interest expense (crores)	243	267	324
Tax (crores)	70	127	189
Net profit margin (%)	0.8	2.9	4.4

Source: Care Rating Report (2019).

4.2 GROUND REALITIES, RELATED ISSUES, AND CHALLENGES

4.2.1 FRAGMENTED MARKET AND LOW ENTRY BARRIERS—RESULTING IN INTENSE COMPLETION ALL AROUND

As discussed in the earlier chapters, the Indian food and beverage services market has many small and mid-size unorganized players competing with a wide range of small and large, chain/nonchain operators (domestic as well as Indian brands) offering products across various formats (cafes, QSRs (quick service restaurants), casual dining, fine dining, etc.). This fragmented market represents a number of challenges, including sometimes unclear format segmentation.

Customers today have a wide array of options in the form of multiple food and beverage operations willing to offer products and services that are more or less similar at competitive prices. The depth and width of the competition for each format (organized or otherwise) ensure that no single player usually leads the market; and any leadership position achieved by a brand is challenged quickly as sustainable competitive advantage is hard to both achieve and retain. Consumer loyalty exists at some level but is subject to a value proposition being maintained (Bowden et al., 2013). It is challenging for any business to attract and retain customers, brands like Pizza Hut and Dominos have a large market share between them, but even in their case, threat of local brands offering products at cheaper prices is quite real and thus even for well-entrenched players a fair level of performance-related uncertainty is to be expected. Given the low barriers to new entrants, it is critical that a food-service business is always reinventing itself, creating excitement around the food and beverages on offer, evaluating the service standards and ensuring that the targeted customer base is diverse enough to allow the business to grow and remain hedged against overdependence on any specific customer segment.

The above discussion however must also be evaluated for relevance across various food and beverage formats and their inherent focus on certain particular socioeconomic–cultural market segments, while some formats are more emendable toward diverse customer segments (QSR, CDR (casual dining restaurant), and cafes) others like (FDR and bar cafes) are limited in that respect for obvious reasons. The landscape is going to remain so in near to midterm; thus, the semi-urban, rural India remains the critical geographic region for organized formats (chain or otherwise) to grow; that said, the unorganized business will remain significant and a factor that will affect how the organized segment expands its market share in India.

4.3 SYSTEMIC AND MACROENVIRONMENTAL ASPECTS TO CONSIDER

4.3.1 MANPOWER CHALLENGE: SHORTAGE OF WELL-TRAINED STAFF AND RETENTION

The Indian hospitality industry is highly labor-intensive, availability of trained, qualified chefs, managerial staff, and other staff for the line positions are often stated by industry leaders as one of the key factors limiting their growth. The ministry of skill development and entrepreneurship in its annual report 2016–2017 projected employment figures in 2017 for the hospitality and tourism sector at large to be 9.7 million and also mentioned that the numbers are expected to reach 14.6 million by 2022, implying an incremental human resource requirement of 4.9 million by 2022 (India Skills Report, 2017). As stated earlier the challenge clearly is to find employable manpower to fill vacant positions across a wide range of job roles. The national skills development corporation (NSDC, 2017) in its report on human resource and skill requirements for the hospitality sector for 2020 identifies the gap for the sector as a whole and can be interpreted to be equally relevant for the food and beverage services segment of the sector as well. Over the last decade or so as the food and beverage space has expanded, so has its share in the employment block as reflected in Table 4.2.

TABLE 4.2 Share of Manpower in the Hospitality Industry in India.

	Hotel (%)	Travel and tour (%)	Restaurants (%)
2016	6	24	70
2021	5	18	77

Source: NSDC Report (2017).

The data allow one to recognize that it is the restaurants where more than two-third of jobs exist. Another issue that deserves mention is the high attrition rate of 20–25% that most operators in the industry have to deal with. It is actually a vicious circle of sorts as shortage triggers poaching (often for small increase in payout) and lack of stability across all job positions inherently harms everyone, attrition upward of 25% is not a surprise to any operator in the food and beverage industry.

To bridge the demand and supply gap businesses end up hiring warm bodies and then invest in training them to get them shop floor ready, through in-house training programs or external training providers. The National Skills Development Corporation (NSDC) under the ministry of skill development and entrepreneurship recognizes this and has been set up as the apex body to support the industry. As a part of that mandate the NSDC spearheaded the National Skills Mission, set up sector-specific skills councils to build the required infrastructure and approach required to get interested participants trained (operational skill areas of interest) for free (or very nominal fee) through short-term training programs offered by NSDC partners. The initiative was aimed at ensuring that employment opportunities were generated for those people who were unemployed and also supporting the industry with its chronic manpower challenge. Despite almost five years, the response is at best mixed despite huge financial outlays being made by the central government.

There seem to be multiple reasons for why the demand–supply gap has not been resolved; this is despite government-led initiatives both at the central and state-government level. The rural and semi-urban areas in general and the lower income social groups have been the target segment, these are the very groups where English as a medium of communication comes with a big question mark, rendering these aspirants not fit for employment by the top-tier hospitality, food, and beverage service businesses, the budget segments are able to offer opportunities, but the remuneration offered is not quite competitive enough to attract the job aspirant to hospitality businesses. To top it off long hours, poor work–life balance all play a role in further dissuading the potential candidates to the job pool to seek other pastures, the result is that year after year the industry continues to find shortfall of employable talent as a key issue to deal with. Tiwari and Munjal (2019) make recommendations that can help leverage the Skill India Initiative to support the industry in meeting its labor requirements:

- School education must focus on building employability related skills (critical thinking, communication, problem-solving, and interpersonal skills).
- Effective collaboration between academic/training institutions and industry professionals is much needed to remove wrong notions and generate goodwill required to encourage young job seekers to the profession.

- The English language gap is proving to be a critical bottleneck. Finding a solution to this challenge is going to be the key to the recruitment of candidates certified under "Skill India."

They also stated that at the broader policy level the government both at state and central level should focus on enabling job-oriented training programs based on industry expectations, issues, and challenges faced by all stakeholders should be responded to in a timely manner as very often stakeholders are frustrated by the red tape that comes with anything that the government tries to implement. The industry can also play a stronger role by allocating more resources toward skilling initiatives either as a corporate social responsibility mandate or otherwise.

The food and beverage services business employ a large percentage of the overall manpower deployment as mentioned earlier and thus must recognize the need to build an ecosystem with ample new entrants joining the job pool rather than having businesses compete for the same stagnant pool.

4.3.2 INFLATION AND IMPACT ON INPUT COSTS

Food price inflation affects the food and beverage services market in India as it is significant "direct cost" for nearly every F&B format, depending on the type of format under consideration the cost of "food and beverages" may constitute anywhere between a 30% and 45% of the total expense spread of a typical P&L statement. The fast food, QSR, and casual dining formats would do lower food cost numbers in percentage terms (based on their menu offerings, quality of ingredients, and use of frozen/processed raw materials) as against the premium dining segments. What makes the raw material cost an important aspect impacting the profitability of the food and beverage business is not just the significant chunk that it represents as an expense head but also the fact that the remaining cost heads be it labor, rent, and utilities are for most part not amenable to variability and behave like fixed costs. This makes material cost a critical factor in business profitability.

Raw material cost (food cost) can be impacted by multiple factors like delayed or deficient rainfall during monsoons, broader economy-related aspects like a slowdown reflected by poor GDP growth, and any unfavorable demand–supply conditions. Food inflation fluctuates and depending on factor conditions mentioned earlier and affects the cost of food-related raw materials on an ongoing basis. If we look at just the last decade, we will observe that food inflation reached a high of 18% in 2010 and then tapered off, but remained in double digits for quite some time (Statisticstimes.com, 2018). This aspect affects consumer indulgence across all food and beverage formats and impacts the profit margins of operators directly. Any perpetual rise in food costs adversely impacts the margins, compelling them to either increase menu prices or find other solutions to mitigate the profit erosion. Increasing menu item prices are seemingly an easy enough choice but in reality in a highly competitive market like India where the challenge of retaining customers who are already value conscious is quite strong, increasing prices may not always be an option on the table, the operators are often forced to look for other innovative ways.

Munjal and Sharma (2011) in their research on the response of the budget food-service restaurant segment in India toward inflation in food-related raw materials found that businesses chose either a strategy focused on cost reduction or one focused on pricing to handle the revenue dilution, some were we found to respond to the challenge by following a mixed approach as well. Irrespective of the food and beverage segment, it is apparent that the approaches discussed below find a general relevance across formats.

4.3.2.1 FOOD COST REDUCTION CENTRIC APPROACH

The operators recognized the need to manage their input costs by working toward using the internal capabilities to reduce food cost by adopting process and/or product changes. In case of those operating in budget/economy market segments, the price increase opportunity was minimal and hence this strategic choice found favor. Even those F&B formats like the fine dining, where price sensitivity was seen as less of a factor the need to be cost-conscious to remain competitive was not lost on the management.

4.3.2.1.1 Innovative Solutions and Practices Aimed at Building Cost-Efficiencies

1. *Purchase contract renegotiation*: Negotiating for better rates, buying in bulk at lower rates, purchasing internationally where

rate differential exists even after factoring in transportation costs, building strategic partnerships that allow common purchase mechanisms being shared by multiple participating food and beverage operations are some of the approaches that can help get lower prices from suppliers.

2. *Aggressive yield management practices*: Effort to improve practices through staff training to improve yields, another approach included unique uses of otherwise wasted raw material components, for example, use of orange peels to prepare a flavorful welcome drink or a mocktail.
3. Menu modifications, menu balancing, special menu/chef promotions are used as tools to respond to the food inflation challenge—food and beverage operations—often offer new menu items where the inflationary impact is lower, or the vendor contract was protective in terms of rates charged. Special menus find play depending on how the chefs are able to put their skills and knowledge at work to come up with dishes with stronger contribution margins. Recipes using cheaper ingredients can be experimented with to evaluate options.
4. *Portion size reductions*: This becomes a typical strategy for operators on the ground where price increase is not an option, and the menu selections could not be taken off, given the current trend around "less is more" portion size reductions are being well received by the customers in the fine-dining segment, chefs focus on creativity in food presentation and offer personalized service to drive the value proposition in the eyes of the customer.
5. *Increased focus on TDH (table d'hôte) menus/buffets/combo meals versus Ala Carte menus*: Operations across multiple dining formats have been able to manage food cost better by offering fixed menus mentioned above; this approach also helps manage labor cost more effectively. Some of the other benefits include faster cover turnover, revenue maximization. Chefs and restaurant managers across formats have developed and offered menus innovatively to lower food cost and also experiment with ingredients.
6. *Analytics driving focus on high margin menu items*: It is never been more important to analyze the data offered by point of sale (POS) applications akin to menu engineering concepts that were taught and forgotten. Not anymore, restaurant operators are analyzing the

sales data to understand about SKUs (stock keeping units) that make them money and those that do not. Allowing them to use the analysis to then-train service teams to push sale of items with stronger contribution toward business profitability and strengthen their menus with items where the margin of profit was high. This where technology is helping even smaller businesses, POS solutions like "Rista" and many other cheaper IT solutions are now available, high-cost providers like Macros, and so on, that were unaffordable for smaller businesses too will need to evaluate their value proposition given the number of competitors in the IT space.

7. *Backend integration of the supply chain*: As focus on sustainable practices has become a mantra of sorts, many food-service businesses have latched on to a back end integration drive by setting up kitchen/vegetable gardens or cultivation efforts that allow them to source some of their produce needs and in some cases even dairy needs. Although the economics behind these efforts may not be quite what the businesses may be expecting, most of those who have made efforts and investments in this direction have used the same to source fresh produce that is often organically produced as well as a USP (unique selling proposition) to market themselves and even take a price hike given that they are using farm-fresh produce (Munjal et al., 2016).

4.3.2.2 MENU PRICING CENTRIC APPROACH

Though not for all formats and market segments, the option to protect profit margins affected by higher input costs resulting from inflation through price increase can and is exercised. It works better and with limited reciprocal impact for formats that are servicing market segments that are less price sensitive. As a strategy a price increase that is offset by revenue correction due to loss of customers is not usually workable, many a times an operation has to test the price elasticity to get a more realistic view of the flexibility in pricing that the business has at its disposal.

It is not uncommon for food and beverage operations to deploy a mixed approach toward handling the profitability erosion caused by higher input costs driven by inflation, they factor in the impact of cost reduction strategies and tactics deployed and supplement the same with marginal price increases such that the combined impact allows them to achieve their profitability objectives without losing the customer base.

Earlier chapters have already articulated the immensely competitive nature of food and beverage business in India. The macroeconomic pressures caused by a weakening economy with the problem of food inflation only make it an even harder situation for the restaurant business. The last decade has seen Indian economy grow (real GDP (unique selling proposition) terms) at a high of 10.3% in 2010 to low of 3.9% in 2008, 2015 onward the economy seems to be on slower growth trajectory, while the overall inflation has not been an area of concern more recently, but with a build-up around agriculture sector slow down, weak monsoon with draught like conditions the food-related inflation can run out of control fairly quickly and the food-service industry would do itself a favor by having a plan of action ready to grapple with any such adverse scenario. Given the need to support the small farmers with loan waivers, minimum support prices and a slew of other subsidies on seeds, fertilizers, and pesticides, it is apparent that the agricultural ecosystem is in distress, climatic changes, and natural disasters to have had their say in the worsening agricultural situation. All in all the issue is likely to remain a flashpoint for the sector in terms of a strategic response. Table 4.3 provides a guidance how the GDP and Inflation as macroeconomic indicators have played out over the last decade.

TABLE 4.3 GDP and Inflation Statistics for India.

Year	GDP growth (%)	Inflation (%)
2008	3.9	9.1
2009	8.5	11.0
2010	10.3	9.5
2011	6.6	9.5
2012	5.5	10.0
2013	6.4	9.4
2014	7.4	5.8
2015	8.0	4.9
2016	8.2	4.5
2017	7.2	3.6
2018	7.1	3.5

Source: Statisticstimes.com (2019).

4.3.3 REAL ESTATE AND LAND REFORM ISSUES

Food and beverage businesses have always recognized "location" as a critical success factor; it is often quoted that getting the location right is the single most important predictor of business success. Rising cost of commercial space rentals, leases, or even owned options as compared to international standards is turning out to be a business dampener for food and beverage businesses in India.

For over a decade, India has been experiencing an escalation in real estate prices amplified by increasing demand and the availability of easy credit. For food and beverage service operations, real estate cost (rentals) is typically the second major cost component after raw materials and can account for anything between 10 and 15% of the business topline and sometimes even higher than 15%.

High real estate rental costs are further made worse as owners are subjected to high services tax on property sale–purchase, despite claims of tax reforms, the land and real estate related taxation has still remained largely ignored. Laws around land availability, land ceilings, and floor space index (FSI) remain regressive, governmental red tape continues to cost developers through delays in approvals as much as corruption around getting illegal approvals by circumventing bylaws and regulations. The permissible FSI in India varies across states and cities and, on average, is in the range of 1–4. In comparison, the global FSI average is 10. The regulations are often guided by larger issues like population density and carry capacity of already under pressure urban cities, For example, in Mumbai, where space shortage is an issue; the FSI (including transfer of development rights) is 2. Given the ground reality and market dynamics, the businesses often respond through unauthorized constructions that are under constant threat of punitive action in the form of fines, demolition drives (to bring down illegal constructions), and bribes to concerned government officials to keep them from taking action.

The real estate cost (rentals or otherwise) is a larger challenge in the urban India, semi-urban, rural India it is not that much of an issue, but with economic growth and development the tier 2, 3 cities and semiurban locations too are headed that way. Delivery based formats and "cloud kitchens" are emerging as innovative solutions allowing food and beverage businesses to exist sans the real estate cost to a significant extent, but then many a format will continue to need the space, esthetics, decor in right location to create dining experiences for customers. The large unorganized segment too has implications from the real estate cost side of things but again at a diluted level. The *Real Estate (Regulation and Development) Act 2016* is a forward-looking regulation that was much needed, it has potential to regulate the flow of black money, hold developers responsible, and protect the interest of investors at large, the early impact of the act coming into place is the across board rate correction much to the relief of both commercial and residential real estate owners/renters alike. The Economic Times (2018) reports JLL guidance on a more than 80% uptick in demand for retail space in malls, indicative of a strong emerging demand. This suggests that the price correction on the back of economic recovery may be spurring a demand recovery of sorts.

4.3.4 FRAGMENTED SUPPLY CHAIN, STRUGGLING COLD CHAIN, AND THE LOGISTICS NIGHTMARE

The food-service industry in India has been adversely affected by a supply chain that is fragmented in nature and marked by the presence of multiple intermediaries. The lack of appropriate transport and storage infrastructure, inadequate technologies, and nonintegration of the food value chain (including the cold chain) are recognized as factors resulting in nearly 30–40% food wastage across various parts of the supply chain. The food cost impact due to inflation that was discussed earlier is further aggravated due to supply chain-related losses, making it essential for food-service operators to recognize the need to identify existing supply chain issues and work with the government and private investors to implement suitable countermeasures to stem the rot.

- Storage (dry and cold) and Warehousing: There is a severe a shortage of quality warehousing space in India. More often than not the available warehouses are more like sheds; they lack the optimal size, design, and temperature-controlled storage system and are nowhere near international standards. The available warehousing space is clearly inadequate with respect to the growing demand. According to industry sources, the total warehousing space in India is estimated at 80–100 million metric tons (MT), whereas the demand is for a much larger space. Currently, cold chain facilities in India are inadequate to meet the prevailing demand. The existing cold chain infrastructure is focused on storage (85%), whereas other 15% for transportation but again the quality of this infrastructure is quite dismal.

By industry estimates, there are about 6000 odd cold storage units in India with a total installed capacity of nearly 25 million MT. Around 25% of these cold storages are multipurpose, with the rest for storage of potatoes (nearly 55%), other fruits and vegetables, dairy products, and meat and fish. In terms of storage capacity, with nearly 80% space being utilized for potatoes, only 20% space is left available for other commodities. Notwithstanding the larger supply chain issues discussed earlier, the need for enhanced cold storage infrastructure and advanced temperature-controlled storage solutions (also during transit) is critical to minimize wastage of fresh produce, extend life of perishables, and enhance quality of food raw materials that the food-service industry requires.

Transport and logistics: The logistics market in India is highly unorganized with only a few national level operators. Despite some efforts by the recent government, the infrastructure is still largely inadequate to meet the demands of a growing industry. Roads, rail network, and ports' infrastructure all need investments from state and central government, and where possible a push toward privatization. Insufficient distribution channels, overburdened ports, and poor quality of logistical services are real bottlenecks that impact food-service businesses in the Indian market. The fact that logistics-related costs constitute over 13% of India's GDP as against 5% in developed countries further establishes the need to recognize the challenge this issue presents. Given the focus and growth in food delivery format, there is an immediate need for professional delivery/logistics as timely delivery of food keeping all HACCP (Hazard Analysis Critical Control Points) aspects in mind is the key to customer satisfaction. Table 4.4 identifies some prominent players in the market.

TABLE 4.4 Key Players in the Logistics Segment.

Company	Inception	Operations	Funding (till April 2016)
Shadowfax	2015	6 cities	USD 8.8 million
Roadrunnr	2015	11 cities	USD 21 million
Delhivery	2011	Pan India	USD 127.5 million
Scootsy	2015	Mumbai	USD 2 million

Source: NRAI (2016).

The need of the hour is to establish an end-to-end supply chain, making use of modern structures such as logistics parks, integrated cold chain solutions, and last-mile connectivity, and also adopting such technology as barcoding and RFIDs (radio frequency identification) and assuring government support through the incentivizing of private enterprises.

4.3.5 LIQUOR SOURCING, RETAILING, AND CONSUMPTION RELATED ISSUES

When it comes to all aspects around sourcing, sale, and consumption of liquor, the sociocultural dynamics come into play, as discussed earlier in the book, there is an overall negative disposition on the issue on one hand, and diabolically there is a sharp growth in consumption (demand side) indicating defiance despite odds. Complex and often vestigial regulations exist at both the central and state-government level when it comes to sourcing liquor. A food and beverage operator offering alcoholic beverages has to obtain a series of licenses to sell alcohol and also adhere to often archaic rules around permitted hours, eligible age limit for consumption, and so on. Some other linked issues are:

- Long gestation period: It takes anywhere between 1 month to 1 year to get a liquor license, with loads of paperwork, and approvals from multiple departments. Lack of transparency often opens up an opportunity for consultants to offer services in getting the job done; their charges often include payouts to be made to government departments, in addition to the charges for their services.
- Antiliquor mindset: Many states have been "dry" historically, some turned dry recently, the politics of the region affects the antiliquor mindset. Liquor procurement is discouraged in some states by applying simulated restrictions such as high tax rates, monopolizing procurement, and levying charges on distribution.
- Spiraling Licensing fees: An estimated INR 0.5–0.6 crore (USD 0.10–0.12 million) is typically required by a restaurant-bar operator to fulfill all licensing requirements, which is a rather large capital outlay, often impacting a restaurant's ability to remain profitable. In addition to the onetime charge, the reoccurring annual license fee to sell imported alcoholic beverages is also quite high, for example, this fee is INR 10 lakhs (USD 19,230) per year in New

Delhi. Processing charges again are additional and most operators end up hiring consultants to manage these requirements, incurring more expenses.

As sociocultural change especially in urban India drives consumers to accept social drinking as a way of life, the options before the restaurant-bar operators are to cough up the dough and get required licenses to purchase and sell liquor or find fringe options like those represented by "BYOB" concept that was discussed earlier. Not being able to sell liquor puts a food and beverage outlet at a disadvantage, sometimes resulting in it's not being able to survive, any push toward rationalization of the mentioned costs and transparent processing would be a big help to the industry.

4.3.6 HIGH-TAX BURDEN FOR THE ORGANIZED FOOD AND BEVERAGE SERVICE SEGMENT

Tax reforms in India seem to have given the food and beverage service sector a miss. Direct taxation in the form of "goods and services" tax (GST) was expected to create a level playing field across the states for food and beverage service businesses, provide relief to customers; however, the regulation started with even more complex structure, separating "air conditioned" and "nonair conditioned" service locations with different tax slabs. Over the last couple of years through lobbying with the government the industry has got some concessions, yet issues around high "VAT" (value-added tax) on liquor and implementation of service charge remain unresolved.

Indirect taxes like customs duties, excise permits, and additional fees like government education surcharge are levied on a range of imports and especially on the import of alcohol. For example, a 150% tariff a levied on imported wine upon which the central government further imposes an "extra additional duty" (EAD) of 4%. The ordeal does not end here as the state governments are free to levy their own taxes on alcohol that range from 30% to over 100%.

In essence the Indian restaurant industry is burdened with multiple taxes like GST, VAT, excise, and services charge (not a tax), besides different state taxes, the direct taxes typically add up to 17.5–25% of the bill value that a customer pays, this tax is much higher in comparison to other sectors like apparel and footwear, where it ranges from 10 to 15%, implying a need for the government both at center and state to review the tax structure at a holistic level. Taxes are usually passed on by restaurant operators to consumers, as a result based on spending power; customers calibrate the frequency of their visits to restaurants. Overall high taxation is a serious impediment to market growth. Table 4.5 provides a list of taxes that a food-service operation needs to deal with (NRAI Report, 2016).

TABLE 4.5 Taxation in Food and Beverage Services Outlet.

Nonalcoholic beverages	Under GST
VAT—alcoholic beverages	18% up to 35% (varies from state to state)
Goods and Services Tax (GST) has two components: CGST—central government part, SGST—state-government part	• 5% GST on food services provided by restaurants (both air-conditioned and nonair-conditioned) • 5% GST on restaurant services including room service and takeaway provided by restaurants located within a hotel featuring room tariff less than Rs. 7500 • 5% GST on any food/drink (nonalcoholic) served at cafeteria/canteen/mess operating on contract basis in office, industrial unit, school, college, hostel, etc. • 5% GST on meals/food services provided by Indian Railways/IRCTC or their licensees both onboard trains and on platforms • 18% GST on restaurant services including room service and takeaway provided by restaurants located within a hotel featuring room tariff over Rs. 7500 • 18% GST on food services including delivery of food provided by a restaurant/food joint located within premises of a club, guesthouse, etc. • 18% GST applicable to all outdoor catering services provided
Service charge	This is not a tax but an expression of gratuity, customer can refuse to pay by law, varies from 5% to 10%.

Source: NRAI (2016).

4.3.7 *OUTDATED LICENSING LAWS AND JUDICIAL OVERREACH*

4.3.7.1 OVERLICENSING

In India, obtaining the requisite licenses, e.g. health license, food safety license, police license, "No Objection Certificate" (NOC), from the fire department, the state pollution control board, and so on is a major challenge for any investor getting in the business, and often becomes a key barrier to entry of competition and a deterrent to smooth operations of a restaurant. Despite government claims the process is neither transparent or centralized (single-window clearance) as yet and requires filing applications with multiple individual stakeholders, involving a lot of paperwork that is often confusing and time-consuming, resulting in a need to hire consultants who are trained to deal with the various stakeholders. The licenses required to start a restaurant are the same throughout India, expect in some states like Maharashtra. An operator typically needs approximately 12–15 licenses just to open a restaurant from various government departments. When this requirement of multiple licenses is compared with international standards one wonders what is so hard to fix in this regards, why is there no shift in policy or even judicial action to push the legislature to wake up to business challenges of the day.

No surprise that big brands usually prefer to form a separate department or hire an external liaison consultant to handle restaurant licensing and other legal compliances because of the complex nature of the job and the time and effort demanded. The laws are also not uniform, differing as they do from state to state, and further, are open to interpretation. Table 4.6 lists the licenses required for opening a restaurant in India. To provide a comparison point, Table 4.7 lists the number of licenses required internationally to open an F&B business, it is quite apparent that we have a clear case of excessive licensing that needs attention (NRAI Report, 2016).

The process of obtaining these licenses is cumbersome and expensive; ad imposes a high and avoidable "cost of compliance" that benefits neither the industry nor the public. The restaurant industry needs to ensure the implementation of a solution to the licensing issue to achieve its full potential, which can perhaps be achieved by encouraging a single-window license clearance.

TABLE 4.6 List of Restaurant Licenses—India.

Food safety license	FSSAI	Mandatory
Health/trade license	Municipal Corporation/ Health Department of Concerned State	Mandatory
Eating house license	Police Commissioner—Licensing	Mandatory in some states
Liquor license	Excise Commissioner	Mandatory for service of liquor in the restaurant otherwise not required
Approval/re-approval of restaurants	Department of Tourism Government of India/State Government	Optional (mandatory for liquor service in Delhi)
Playing of music in restaurants—license	Societies which are registered under Copyright Act by the Government of India	Mandatory when recorded/live music of the Copyright Holders are played in the restaurant
Environment clearance for grease trap/ETP (Water Pollution Act)	State Pollution Control Board	Mandatory
Environment clearance for generators sets (Air Pollution Act)	State Pollution Control Board	Mandatory for gen-sets
NOC from fire department	Fire department of the concerned state	Mandatory
Weights from fire department	Legal Metrology Department	Mandatory
Lift license	Concerned Authority—Electrical Inspector, Office of the Labor Commissioner	If lift is to be installed
Insurance required to be taken: Public liability • Product liability • Fire policy • Building and asset (There are other insurance policies also which are not mandatory but useful)	Any insurance company	Mandatory
Shop and Establishment Act	As prescribed in the Act and as applicable to the state	Mandatory
Signage license	Municipal committee/ corporation of the city	Mandatory in some states

Source: NRAI (2016).

TABLE 4.7 List of Restaurant Licenses—International.

Country	No. of licenses	Nature of licenses
India	12–15	As detailed in Table 4.6
Thailand	5	Sanitation (food) license fire license sales (tax) license work permit license local municipality license
China	4	Sanitation license environment license fire license sales license
Singapore	4	Food/health license fire license sales license waste license
Turkey	2	Municipality license fire department license
USA	7	Business license employer identification no. certificate of occupancy food-service license liquor license

Source: NRAI (2016).

4.3.7.2 JUDICIAL OVERREACH

The ban on sale of liquor by food and beverage service establishments if they were located less than 500 m from any state or national highways was a classic example of courts getting involved with something that was way outside of their role in Indian polity. The ban eventually went away as all stakeholders cried foul. As reported by The Hindustan Times (2017) in a column by Karan Thapar, an eminent journalist, despite the intent, this overreach by the Supreme Court cost the industry dear. At some point, the government, bureaucracy, and the judiciary must recognize the socio-cultural shift consider impact on industry and fine solutions to issues in a consultative manner rather than the kind of knee jerk reactions witnessed so far.

Laws and policies must be aligned with the modern-day life of citizens, be progressive, rather than regressive, without this change the industry will always find itself at loggerheads with regulatory framework that is both irrelevant as well as more of a roadblock to the growth of the industry than otherwise.

4.4 OPPORTUNITIES AND TRENDS TO WATCH OUT FOR

The focus of the chapter so far has been on key issues and challenges that the food and beverage service segment of the hospitality industry in India is facing as it continues to grow as well as evolve. It is not uncommon to witness challenging business scenarios do two things:

- Raise barriers to entry, such that more competition and investment is deterred
- Find the committed players innovate and find opportunities that never existed before

The unorganized format of the food and beverage service business in India has thrived outside of the tax net, with lower input costs (including real estate cost) for decades, the quality of product and services that are offered may or may not be seen as compromised. It really depends on what the consumer is looking for and the overall value proposition that gets chalked up. The roadside "dhabas," "street food" small-time sweet shops are here to stay that is quite easily visible. The cost advantages they enjoy and the demand that exists for what they have to offer is real and not something that organized formats, their professionalism notwithstanding can railroad (Fig. 4.1).

FIGURE 4.1 Street restaurant typical of the unorganized F&B format in India.

Source: Author.

The organized segment of the business has found innovative answers, there are trends that are mentioned below which have come into being as businesses looked for ways to grow and profit despite the issues and challenges. Some of these trends are:

- Cloud kitchens and virtual kitchens: These are essentially production back ends of restaurant and catering businesses minus the front end. The business model allows these operations to be set up at far less expensive locations; the space requirement is only that of the production side of things, these are coming up in a big way, often with online platforms for branding and transactions with a professional delivery system in place (Table 4.8) (NRAI Report, 2016).

TABLE 4.8 Key Players in the Virtual Kitchen Segment.

Company	Inception	Operations	Funding (till April, 2016)
Faasos	2011	15 cities	USD 55 million
Freshmenu	2015	3 cities	USD 22 million
Hola Chef	2014	2 cities	USD 3.42 million
Box 8	2012	3 cities	USD 3.5 million

Source: NRAI (2016).

- Ready to eat and home chef options: Again in more ways than one, the challenges and issues cited earlier in the chapter are the triggers for the growth and popularity of these options. There is a growing demand for "home like," "healthy," "traditional," and "slow food" in urban India. Much akin to a sociocultural shift, a significant number of people are interested in less oily, healthy food cooked with fresh ingredients. The business model that has emerged allows passionate chefs (individuals) to become business owners by becoming part of brands that aggregating their capacity to offer a wide range of products through online platforms supported by professional packaging and delivery. Working professionals, students staying in shared accommodation are fast emerging as a segment interested in this kind of food. Even business travelers staying at guest houses or hotels with limited or no dining services are finding these options quite useful (Tables 4.9 and 4.10) (NRAI Report, 2016).

TABLE 4.9 Key Players in the Ready to Cook/Eat Food Segment.

Company	Inception year	Operations	Funding (till April, 2016)
Innerchef	2015	7 cities	USD 1.66 million
Haute Chef	2015	Mumbai	Not available
Cookfresh	2014	Delhi, Gurgaon	Not available

Source: NRAI (2016).

TABLE 4.10 Key Players in the Home Chef Aggregator Segment.

Company	Inception	Operations	Funding (till April, 2016)
Cyber Chef	2015	Delhi NCR	Data not available
Food Cloud	2013	Delhi NCR	Data not available
Red Cooker	2014	Bengaluru	USD 500k

Source: NRAI (2016).

Technology has had a major role in getting these trends to become commercially viable and thrive, especially the ability to process transactions, payments electronically. Without the technology backbone, none of these trends could have achieved the market penetration and acceptance that has come their way. Table 4.11 lists some of the prominent service providers in the contactless payment space in India (NRAI Report, 2016).

TABLE 4.11 Key Players in Contact Less Payment Space.

Company	Inception	No of merchants	No of wallet users	Funding (till April, 2016)
Paytm	2010	~150,000	More than 120 million	More than USD 775 million
Freecharge	2010	More than 100,000	More than 15 million	USD 113 million (acquired by Snapdeal in April, 2015)
Mobikwik	2009	More than 50,000	More than 15 million	USD ~87 million
Oxygen	2004	More than 9000	More than 20 million	Not available

Source: NRAI (2016).

- **Return of the food trucks, street food, and food festivals:** It is amazing how new trends that are finding interest are in essence a return of the food concepts and their commercial expression that was widely prevalent a couple of decades back. The branding, packaging, and delivery platforms draw from the available infrastructure, especially the digital and technological support systems, but the core of the food-related elements find historical roots with some customization to the current palate alongside a need for a stronger focus on HACCP elements.
  Government too has supported these concepts as potential opportunities for self-employment through financial support that is provided to small and medium-scale enterprises under various government schemes, Skill India initiative has supported training and skilling of those interesting in learning about cooking in a professional way so as to open their own food trucks and kiosks. Many an opportunity to start catering setups or setting up food stalls at food festivals and/or fairs is being enabled as governments at the state and central level recognize the economic potential of tourism-related services/facilities. As mentioned above, these are thriving in both more visible branded organized formats as well as in the form of individual operations that exist as individual small-scale entrepreneurial effort. It is not uncommon to find these entities to exist without any licenses, registration, or even basic billing processes that are essential for tax-reporting purposes. These businesses exist as a part of the gray economy in every sense of the word. The "thela walas" of yesteryears may have given way to seemingly more professional street food vendors often boasting of better infrastructure in the form of "kiosks" or "mobile carts," it is the food that is still at the core of their business, followed by their ability to price the offerings cheap given the lack of overheads mentioned earlier (Table 4.12) (NRAI Report, 2016).

TABLE 4.12 Some Popular Food Truck Brands across India.

Company	Food served	Inception	Operations
M-Theories	Nepalese, rolls, Indian, Middle-Eastern foods, etc.	2013	Ahmedabad
De3	Burgers, hot dogs, pastas, rolls, BBQ wings, etc.	2013	Bengaluru

TABLE 4.12 *(Continued)*

Company	Food served	Inception	Operations
The Lalit Food Truck	Latin American and Lebanese foods	2014	New Delhi, Bengaluru
Frugurpop	Popsicles, tea, cookies, etc.	2014	Gurgaon and Mumbai
Street Food (by Punjab Grill)	Burger, aloo tikki, pav bhaji, kullad chai, masala chach, etc.	2014	Gurgaon
Drifters Café	Chinese and Pan Asian	2015	Gurgaon
EGG Jactly	Egg dishes, Belgian waffles, Maggi, pasta, etc.	2015	Delhi-NCR

Source: NRAI (2016).

In addition to the above, more professionally managed catering outfits, often eventually evolving into event management companies are becoming popular. Although events management business is a different domain altogether with further layers of specialization like those who focus only on weddings, destination weddings, sporting events, and so on, food remains an integral part of what these operators have to offer. As a result as the events industry has grown and gotten more specialized, so has the need for more contemporary food and beverage products and services (Table 4.13) (NRAI Report, 2016).

TABLE 4.13 Popular Food Festivals.

Name of food festivals	Place	Month in which generally organized
Grub fest	New Delhi, Gurgaon, Pune	March, October, December
Asian Hawker Market	New Delhi	February
Gourmet High Street	Gurgaon	December
Palate Fest	New Delhi, Goa	November
Family Food Fest	Mumbai	December
Bengaluru Food Fete	Bengaluru	April
Ahare Bangla	Kolkata	October–November
Street Food Festival	Chennai	February

Source: NRAI (2016).

- **Upmarket Food Hubs:** In an era of mixed-use real-estate development, integrating the food and beverage operations in a manner where the daytime customers are the office goers (working out of the commercial/office spaces) and the same setting becomes the evening hotspot for the people living in the surrounding residential areas, alongside the office goers heading out is a trend to watch out for. The cyberhub in Gurgaon is one such example, with dozens of food and beverage service operators offering their products and services to a wide spectrum of customers, serving all three meals. All formats discussed in the earlier chapters coexist in a unique setting. Although this is urban phenomenon for now, it is only a matter of time that the model finds acceptance in tier 2 and 3 cities as well.
- **Travel Infrastructure expansion driven F&B opportunity:** Governments at both center and state level have in recent years made huge outlays to develop travel-related infrastructure in the form of new airports or modernization and expansion of existing ones, big ticket highway projects aimed at building expressways to connect all parts of the country, but especially the key tourist attractions, and even better rail infrastructure in the form of not just modern trains but a lot of investment on railway station infrastructure, including the services and facilities that are made available to the travelers.

This has opened up a huge opportunity for the organized food and beverage service segment in India, akin to developed countries in the west, "grab n go," "vending" as well as retail formats are finding business opportunities. There are examples of roadside eateries (dhabas) developing into larger, more professionally managed F&B operations serving thousands of travelers, Murthal on way along national highway 1 (popularly known as the GT Karnal Road) is one such example.

All modernized airports boast of dozens of F&B retail options, and railway stations are next in line as the government lines up investment. There are issues around getting licenses and the usual red tape where the government is involved, but as push for private investment continues alongside transparency in how the concerned government departments function; there is humungous potential for nearly every food and beverage service format to find play.

4.5 CONCLUSION

As far as trends and opportunities go, there are plenty of cuisine related shifts as well on the ground both in terms of the food as well as beverage domain. That aspect is not being discussed here and would be looked at separately in the book.

Indian food and beverage service industry is a dynamic as well as happening space with so much coming through in terms of new trends, market shifts, changes in customer preferences, and buying behavior. The issues and challenges that the chapter highlighted and discussed are very much the reality of the day, those doing business in India currently as well as those looking toward in India to invest in this space need to be mindful of these aspects. The journey of India's economic liberalization has been two steps forward and one backward, this hesitancy possibly stems from the discord between the rural–semi-urban India and urban India with their unique, often opposite expectations from the government. This holds true for the F&B service industry as well.

It would have become apparent to readers that there are many India's within India. The diversity that manifests itself socioculturally is also manifested in sharp inequalities which will at times make it impossible for powers that be to ignore. Being sensitive to this uniqueness of India will be useful as businesses try to find growth and profitability in the unique market that India presents.

KEYWORDS

- **F&B service**
- **growth opportunity**
- **challenges for F&B service industry**
- **F&B trends in India**
- **Indian F&B**
- **F&B service in India**

REFERENCES

AF and Company Consulting IBEF.ORG, India Brand Equity Foundation. Hospitality Trend Report, AF and Company Consulting IBEF.ORG, India Brand Equity Foundation,

2018, [Online]. https://www.ibef.org/download/Tourism-and-Hospitality-January-2018.pdf (accessed April 20, 2018).
Annual Report (2016–17). Ministry of Skill Development and Entrepreneurship, 2017. Available at: www.skilldevelopment.gov.in/assets/images/annual%20report/Annual%20Report%202016-2017%20-%20English.pdf (accessed January 20, 2018).
Annual Report (2017–18). Ministry of Tourism, Government of India, 2018.
Bharadwaj, S. G.; Varadarajan, P. R.; Fahy, J. Sustainable Competitive Advantage in Service Industries: A Conceptual Model and Research Propositions. *J. Mark.* **1993,** *57* (4), 83–99.
Care Ratings. Indian Restaurant & Food Service Industry—Structure & Prospects, 2019 [Online]. http://www.careratings.com/upload/NewsFiles/Studies/Restaurants%20%20QSRs%20May%202019.pdf (accessed July 10, 2019).
Engel, J. F.; Blackwell, R. D. *Consumer Behavior*; Dryden Presee: New York, NY, 1982.
Euromonitor. Euromonitor Report on Cafes and Bars in India, 2019.
Euromonitor. Full Service Restaurants in India, 2015 [Online]. http://www.euromonitor.com/.
Federation of Hotel and Restaurant Association of India. Indian Hotel Industry Survey. Federation of Hotel and Restaurant Association of India in Consultation with Hotelivate, 2018.
FICCI. FICCI Report: Unlocking the Potential in Food and Beverage Services Sector, 2015 [Online]. http://ficci.in/spdocument/20589/F&B-Report.pdf (accessed June 12, 2018).
Food Service India. Key Trends Impacting Food Service Industry in India, 2019 [Online]. https://www.indiaretailing.com/2019/01/17/food/food-service/expert-speak-key-trends-impacting-foodservice-industry-in-india/ (accessed May 10, 2019).
HVS-Anarock. India Hospitality Industry Review. HVS-Anarock. 2018, [Online]. https://www.hvs.com/article/8488-hvs-anarock-india-hospitality-industry-review-2018 (accessed April 8, 2019).
IBEF.ORG. India Brand Equity Foundation, 2017a [Online]. https://www.ibef.org/download/Tourism-and-Hospitality-January-2017.pdf (accessed April 5, 2018).
IBEF.ORG, India Brand Equity Foundation, 2017b, [Online]. https://www.ibef.org/industry/retail-india.aspx (accessed January 4, 2019).
IMARC Group.com. India Pub, Bar, Café and Lounge (PBCL) Market: Industry Trends, Share, Size, Growth, Opportunity and Forecast 2019–2024, 2018 [Online]. https://www.imarcgroup.com/india-pub-bar-cafe-lounge-market (accessed January 10, 2019).
India Skills Report. United Nations Development Program, 2017 [Online]. www.in.undp.org/content/india/en/home/library/poverty/india-skills-report-2017.html (accessed January 20, 2018).
India Skills Report. United Nations Development Program, 2018 [Online]. www.in.undp.org/content/india/en/home/library/poverty/india-skills-report-2018.html (accessed January 20, 2018).
JLL, India Hospitality Review, 2017 [Online]. http://www.jll.co.in/india/en-gb/Research/Ready-steady-go-India-hospitality-industry-review-2017.pdf (accessed May 10, 2018).
Munjal, S.; Bhushan, S. The Indian Hospitality Industry: Dynamics and Future Trends; Apple Academic Press: New Jersey, USA, 2017.

National Skills Development Corporation Report (NSDC). Human resource Requirement in the Travel, Tourism and Hospitality Sector, 2017. www.nsdcindia.org/sites/default/files/Travel-Tourism-and-Hospitality.pdf (accessed January 20, 2018).

NRAI. India Food Services Report, 2016. National Restaurant Association of India.

Restaurant India.in. Issues and Challenges faced by Restaurant Industry, 2013 [Online]. https://www.restaurantindia.in/article/Issues-and-Challenges-by-Restaurant-Industry.6121 (accessed May 20, 2018).

Munjal, S.; Tiwari, S. Skill India Initiative—Implications for the Hospitality Sector in India. *Worldwide Hosp. Tour. Themes* **2019,** *11* (1), 4–9. https://doi.org/10.1108/WHATT-10-2018-0062.

Tiwari, S. Munjal, S. What Are the Key Challenges That the Indian Hospitality Industry Is Facing in Search of Skilled Labour?. *Worldwide Hosp. Tour. Themes* **2019,** *11* (1), 99–102, https://doi.org/10.1108/WHATT-10-2018-0063.

Statisticstimes.com. GDP Growth of India, 2018 [Online]. http://statisticstimes.com/economy/gdp-growth-of-india.php (accessed January 10, 2019).

The Economic Times. Retail Space Demand in Malls May Rise 81%—JLL. *The Economic Times*, 2018 [Online], https://economictimes.indiatimes.com/industry/services/property-/-cstruction/retail-space-demand-in-malls-may-rise-81-this-year-jll/articleshow/62931225.cms?from=mdr (accessed May 10, 2018).

The Hindustan Times. Supreme Court Order on Alcohol Sale a Case of Judicial Overreach. *The Hindustan Times*, 2017 [Online]. https://www.hindustantimes.com/columns/is-the-supreme-court-order-on-alcohol-sale-a-case-of-judicial-overreach/story-MTsTqrtw5IF38jfEZEsI3I.html (accessed January 12, 2019).

CHAPTER 5

Emerging F&B Business Models in India: Technology Disruption and More

ABSTRACT

The growth story of the Indian food and beverage service industry would not have come thus far without substantial investments being made. The chapter looks at the various sources and routes that these investments have come from and the business models that are at play. Domestic and international 'funds' and 'private equity' and 'venture capital' have been prominent sources, with FDI (foreign direct investment) norms being eased. International brands have been able to invest and set shop in India. While a lot of chain-based F&B formats have followed the franchise route, other models too have been successful. The chapter discusses how technology seems to be at the core of changes and innovation in the industry and is also attracting investment toward foodtech ventures.

5.1 INTRODUCTION

The growth of the varied food and beverage formats in the organized space has been driven by multiple types of business models as well as investment sources. This is in sharp contrast to how the still dominant unorganized F&B space operates from a perspective of investment and approach toward growth.

The unorganized part of the food and beverage service industry has traditionally been quite fragmented and largely driven by personal investments by the owners/operators. Between the available financing options to start and/or grow businesses (mostly family run), SMEs (small scale enterprises) depended on debt in the form of loans from banks or cooperatives. The growth given the limitation of funding options (self or debt) was also rather slow, and only handful operators achieved a status having multiple or chain of food and beverage operations. All other typical aspects of the unorganized format too remain relevant to this limitation. Most of these businesses have and continue to exist without any formal format of organization, a significant number of these (street food, food joints, food carts, sweet shops, etc.) operate without licenses, pay no taxes, and are not even registered as business entities. The transactions are on a cash-only basis and despite the recent push by the governments at the center and state level to broaden the tax net by legislation on the good and services tax system, this group remains a part of the massive gray economy.

Those food and beverage businesses that have achieved scale and some level of professional structuring are usually incorporated as proprietorships or partnerships from a business entity standpoint. This cluster of food and beverage service businesses is likely to be the largest group, still unorganized and fragmented for the most part, especially those in rural and semi-urban India. As discussed in earlier chapters, the transition of the Indian economy from a closed orientation to an open more liberalized economy through reforms that commenced in the early 1990s saw the emergence of what is seen as the "organized" format of food and beverage service business in India. This included a large number of international players making a beginning in India by the introduction of globally popular brands like Pizza Hut, MacDonalds, Subway, and Domino's to name a few. The economic change that was unleashed also saw many Indian companies and brands take initiative to become organized and focus on growth taking advantage of the changing dynamics of doing business in India, as a result soon enough local investment interest and appetite were created. The growth of the organized formats of food and beverage service operations over the last two decades has been strong and was built upon the socio-cultural-economic shift that has impacted all aspects of our life. This growth has been centered around the urban and semi-urban population centers for now, but it is only a matter of time when other parts of the country too witness this evolution. There is no doubt that the unorganized format will remain in play in the foreseeable future, but what is equally true is the shift toward the organized formats is going to continue and is here to stay.

5.2 BUSINESS MODELS DRIVING THE GROWTH

Any business growth will demand investment, as has been the case with the hotel segment of the Indian hospitality industry. The approaches or business models that have largely been deployed globally have also been evidenced in the Indian business space. The hotel brands typically chose to grow either through an "asset heavy" or an "asset light" approach. The long and short of it is about making a strategic choice between expanding without owning all the assets; hence, the use of a range of contracts with typology shared and discussed later in the chapter in detail like management contracts, franchise agreements, etc. or growing slower with ownership of assets.

In the hotel segment, Indian brands like the Indian Hotels Company (IHCL) popularly known as the Taj Group of Hotels and the East India Hotels (EIH) popularly known as the Oberoi Group of Hotels chose the asset heavy strategy for the most part. Even globally hotel brands have done the same; however, it is clear that business models built around the "asset light" model have grown to become market leaders and have also consistently delivered higher profitability. IHG (Intercontinental Hotel Group), Marriott, and Radisson Hotels Group are examples of the success and dominance of the asset light model. In India, more recently brands have made a strategic shift that represents a more middle of the road orientation by also venturing into franchise/management contract based growth strategy.

The food and beverage service space is different from hotels in many ways. The first being the scale of capital investment required, given the range of formats that can vary dramatically, and is likely to be way lower than the capital outlay required to set up a hotel. The expanding commercial space market (real estate) in India, as discussed in earlier chapters, allows lease rental models to thrive, which further dilutes the requirement of capital upfront to get a food and beverage operation started. The lease rental options are available in plenty currently in the metro and tier 1 cities as the supply side outstrips the demand side. A majority of the commercial real estate projects are essentially following a "mixed use" doctrine. The commercial residential use is built in the projects and this too works well for food and beverage service operations as real estate firms are forced to build capacity for a range of food and beverage service formats ensuring that the service requirements for mixed use are met.

Table 5.1 documents the operating models for some of the well-known food and beverage service brands in India. While the majority are international brands, a few of them are local brands. It is clear from the data shared below that prominent business models are:

(a) Master/Multiple/Regional Franchise with an Indian partner(s) who have led the expansion by the addition of more franchise partners.
(b) Ownership format or a mixed approach of owned as well franchise pathway.
(c) Joint ventures with Indian partners to drive both investment and by extension outlet numbers.

It is quite apparent that the "Franchise" approach is the preferred option as far as the organized food and beverage service formats in India are concerned.

TABLE 5.1 Operating Models of Selected International Brands in India.

Brand	Country of origin	Year of entry	Mode of entry	Indian partner	No. of outlets
KFC	USA	1995	Multiple Franchise	Devyani International and Sapphire Foods	352
McDonalds	USA	1996	Master Franchise	Hard Castle Restaurants Pvt. Ltd.	393
			Joint Venture	Connaught Plaza Restaurants Pvt. Ltd.	
Pizza Hut	USA	1996	Multiple Franchise	Devyani International and Sapphire Foods	420
Domino's	USA	1996	Master Franchise	Jubliant Food Works	1039
TGI Friday's	USA	1996	Joint Venture	Bistro Hospitality	15
Café Coffee Day	India	1996	Own	Coffee Day Enterprise	1556
Barista	India	2000	Own	Carination Hospitality	141
Subway	USA	2001	Micro Franchisee	Multiple	568

TABLE 5.1 *(Continued)*

Brand	Country of origin	Year of entry	Mode of entry	Indian partner	No. of outlets
Goli Vada Pav	India	2001	Own + Franchise	Not Applicable	350
Costa Coffee	UK	2005	Master Franchisee	Devyani International	47
Hard Rock Coffee	USA	2006	Master Franchise	JSM Corporation	9
CBTL	USA	2008	Master Franchisee+ Sub Franchisee	Pan India Food Solutions Pvt. Ltd.	33
Chilli's	USA	2009	Regional Franchisee	TexMex Cuisine India Private Ltd and Om Pizza and Eats	19
California Pizza Kitchen	USA	2010	Master Franchise	JSM Corporation	6
Punjab Grill	India	2011	Own+ Franchisee	Lite Bite Foods Pvt. Ltd.	11
Dunkin Donuts	USA	2012	Master Franchise	Jubilant Food Works	74
Starbucks	USA	2012	Joint Venture	Tata Starbucks	81
Pizza Express	UK	2012	Joint Venture	Gourmet Investments	11
Krispy Kreme	USA	2013	Master Franchise	Citymax Hotels India Pvt. Ltd. and Bedrocks Foods Pvt. Ltd.	31
Burger King	USA	2014	Joint Venture	Everstone Group	52
Johnny Rockets	USA	2014	Master Franchise	Prime Gourmet Private Limited	5
Wendy's	USA	2015	Master Franchise	Sierra Nevada Restaurants	4
Carl's Junior	USA	2015	Master Franchisee	Cybiz Corporation	3

Source: NRAI Report, 2016

5.2.1 INVESTMENT APPROACH IN THE F&B SERVICE INDUSTRY

Irrespective of the chosen business model, the growth for food and beverage service brands across a range of formats has to be funded. The expansion has been quite impressive over the last few years and though the data shared earlier are as of 2016, the growth trend seems to be moving forward. As indicated by the table below, the early expansion of most ventures has been funded by either angel investment (self-funded) or some form or other venture capital aimed at getting the business idea off the block. while most start-ups in the food and beverage service industry are still in the growth phase, a few established players have grown significantly and are now in the maturity stage. These have received multiple series of funding from venture capital funds, private equity funds, and the total quantum of funds that have come behind the expansion in India is quite impressive.

A significant amount of foreign investment is also evidenced, as is represented in some of the examples cited in Table 5.2 below. It is only a matter of time as these ventures look at raising capital through the IPO (initial public offering) route from the capital markets by getting listed.

5.2.2 INVESTMENTS BY CORPORATE GROUPS/INDIAN FUNDS

As the growth story of the organized food and beverage service sector in India has progressed and strengthened, the industry has attracted the attention of corporate entities and Indian funds. the table below provides data on a wide range of significant investments mostly in the form of JVs (joint ventures) or equity buy outs. International food and beverage brands have exercised a strategic choice of growing brand presence through partnerships with local players. The strategic call is an offshoot of the overall internationalization approach of global brands wherein they have looked at large markets like India that have huge growth potential as drivers of their own growth given that most developed economies have either slowed down or the market penetration that is almost reached saturation. Partnering with local players brings both much needed local support and connections that are able to resolve entry pangs. Especially, the nature of challenges that India presents has given the licensing requirements and the working of the government machinery at large. While FDI (foreign direct investment) is welcomed in the sector, the growth story so far has seen foreign funds, Indian funds, and corporate entities that invest actively over the last decade.

TABLE 5.2 Stage Wise Investment Approach in the Indian F&B Service Space.

	Early	**Growth**	**Late Stage/Mature**
Type of investor	Angel inventor (friends, family, and private funds of successful entrepreneurs) venture capitalists	Private equity firms	Strategic investors/large private firms
Stage	Start-up stage	Expansion stage	Stable growth and maturity
Operations	Nascent stage where the business idea is in place and the product category defined, although operations may or may not have taken off	The business has already proven concept and achieved reasonable scale. It is in a position to expand aggressively.	Pre-IPO stage when the company has achieved significant scale of operations.
Use of funds	Typically used for setting up restaurants, hiring key personnel, and putting in place systems and processes	Generally used to fund the expansion of current operations of the company	The capital is used to expand the company's operations, enter new markets, finance an acquisition, or provide an exit to the existing investors.
Company size	Small-sized companies with tremendous potential to grow	Mid-sized companies with high growth potential	Large companies with a stable outlook
Size of investment	Small	Medium	Large
Risk profile	Very high due to uncertainty about the business taking of as there is very little or nothing tangible or visible on the ground	Moderate to high; risk is relatively easier to ascertain on the back of existing operations	Lower due to past track record
Return expectations	40–50% p.a.	25–30% p.a.	18–25% p.a.

TABLE 5.2 *(Continued)*

	Early	Growth	Late Stage/Mature
Key examples	Chaayos raised seed funding of ₹ Crore from ZishaanHayath in 2014. In May 2015, the company had eight outlets and raised Series A funding of ₹32 Crore from Tiger Globa.l Holachef raised its Series A funding of ₹19.6 Crore from Kalaari Capital and Mr Ratan Tata in June 2015. After successfully exiting Punjab Grill, Zorawar Kalra started Massive Restaurants with an investment by Mirah Group in December 2012. Swiggy raised Series A funding of ₹12 Crore from Accelindia and SAIF partners in February 2015. Azure Hospitality (Mamagoto) founded by Kabir Suri and Rahul Khanna raised seed funding from GautamThapar (founder and chairman of Avantha Group) and Mauritius based Blue Sky capital.	Barbeque Nation had 12 outlets when it raised funding of ₹110 Crore from CX Partners in March 2013. Devyani International that runs a chain of 180 outlets including KFC, Pizza Hut, and Costa Coffee raised ₹250 Crore in funding from ICICI Venture in May 2011. Cremica raised ₹72 Crore funding from Goldman Sachs in January 2007. It had a turnover of ₹227.5 Crore in FY'07. Azure Hospitality raised ₹66 Crore from Max ventures and industries and Goldman Sachs in May 2016. In July 2015, it raised ₹65 Crore in funding from Goldman Sachs when it had nine outlets of Mamagoto and five outlets each of Rollmall and speedy Chow.	Temasek invested ₹500 Crore in Devyani International (₹400 Crore primary and ₹100 Crore secondry) in October 2014, providing a part exit to ICICI Venture. Devyani operated a chain of over 300 outlets at the time of transaction. Rabo Equity Advisors invested ₹100 Crore in Cremica in June 2016. CX Partners invested ₹390 Crore in Cremica in August 2015 (providing an exit to MotilalOswal PE). In FY'14, in the company recorded revenues of ₹532.5 Crore with an 11% EBITDA margin.

Source: NRAI Report, 2016

TABLE 5.3 Profile of Investments by Indian Corporate and Funds in the F&B Service Space.

Corporate group/Indian funds	Brands	Remarks
Tata Group		Tata Global Beverages and Starbucks Corp. entered into an equal joint venture in 2012.
Bharti Family Office	PIZZA EXPRESS	Entered a 50–50 joint venture to launch the Pizza Express chain of restaurants in India in 2011.
Premji Invest (Azim Premji PE Fund)	Hard Rock CAFE; california PIZZA KITCHEN; 城 shiro	Invested in JSM Corp., a muiltibrand restaurant platform, for a minority stake in 2012
Amtek Auto Group	BARISTA LAVAZZA; Jamie's ITALIAN; Wendy's	Acquired Barista Coffee from Lavazza SPA. It also brought in Jamie's Italian Restaurants and Wendy's to India through master franchise arrangement in 2015
Max Ventures and Industries	mamagoto; SPEEDY CHOW; rollmaal	Acquired a minority stake in the Delhi-based company in 2016
Avan projects (Vinod Mittal Family)	PIZZA PAPA JOHN'S Better Ingredients. Better Pizza.	Invested in Om Pizza and Eats (Indian franchisee of Papa John's Pizza) in 2013
RP Sanjiv Goenka Group	au bon pain café bakery	Entered into a master franchisee agreement with US-based fast bakery and café chain, AU Bon Pain, in 2009

TABLE 5.3 *(Continued)*

Corporate group/Indian funds	Brands	Remarks
Amit Burman & Rohit Aggarwal	LBF Lite Bite Foods	Promoted by the Dabur family, Lite Bite Foods is an F&B retail company with multiple brands under its portfolio
Nandan Nilekani and Rakesh Jhunjhunwala	Café Coffee Day	Invested close to ₹100 Crore in coffee Day Enterprises in its Pre-IPO placement
Aditya Birla PE-Fund	olive	Acquired a minority stake in Olive Bar and Kitchen Pvt. Ltd.
Unilazer Ventures (Family office of Ronnie Screwvala)	mm! maroosh	Acquired a significant minority stake in the Mumbai-based, Lebanese restaurant chain in 2014
Mirah Hospitality	m.o.d; Impresario handmade restaurants; Massive	Acquired 33% stake in Mad Over Donuts (MOD), a donut chain operating in India, in 2010 Invested in Impresario Entertainment and Hospitality in 2011 and Zorawar Kalra's venture, Massive Restaurants, in 2012

Source: NRAI Report, 2016

5.2.3 PRIVATE EQUITY INVESTMENT IN THE FRONT-END SPACE

A wide range of start-ups in the food and beverage space have received private equity investment over the last few years and the trend continues.

The table below offers inputs on a few prominent transactions (NRAI, 2016). It is interesting to note that the bulk of these transactions are in the QSR (quick service restaurant) or cafe formats and a few in the CDR (casual dining restaurant) space. The FDR (fine dining restaurant) format given its comparatively higher capital investment requirement and overall higher risk proposition seem to be getting a pass in terms of attracting private equity investment of the scale that other formats are experiencing.

TABLE 5.4 Private Equity Investments in the Indian F&B Service Space.

S. No.	Target	Key brand(s)	Format(s)	Investor(s)	Amount (INR Crore)	Date	Stake
	Early phase						
1	Charcoal Biryani Restaurants	Charcoal Biryani	QSR	Lion Ventures	1	February 16	NA
2	Thea Kitchen	Biryani Blues	QSR	Carpediem Capital Partners	3	August 15	26.1%
3	Cybiz Brighter Restaurants	Carls Jr India	QSR	Now Capital YouWeCan Ventures	2	August 15	NA
4	Impresa Hospitality Management	Maroosh	QSR	Trans Continental VC	7	August 15	NA
5	White Owl Brewery	White Owl	CDR	RAAY Global Investments	4	June 15	28.6%
6	Chaayos	Chaayos	Café	Tiger Global	32	May 15	32.7%
7	Lazeez Affaire Group	Boombox café café Out of the Box, etc.	Café	Now Capital	NA	January 15	NA
8	Impresa Hospitality Management	Maroosh	QSR	Unilazer Ventures	18	March 14	43.5%
9	Rocket Sandwich	Rocket Sandwich	QSR	Unilazer Ventures, India Quotient	NA	July 13	NA
10	BTB Marketing	The Beer Café	Café	Mayfield	25	May 13	35.0%

TABLE 5.4 *(Continued)*

S. No.	Target	Key brand(s)	Format(s)	Investor(s)	Amount (INR Crore)	Date	Stake
11	Dana Choga	Dana Choga	CDR	SAIF	5	February 13	NA
12	Faasos	Faasos	QSR	Sequoia Capital India	33	January 13	33.0%
	Growth phase						
1	Azure Hospitality	Mamagoto, Speedy Chow and Rollmaal	QSR/ CDR	Goldman Sachs, Max Ventures	66	May 16	22.4%
2	Azure Hospitality	Mamagoto, Speedy Chow and Rollmaal	QSR/ CDR	Goldman Sachs	65	July 15	34.2%
3	Massive	Masala Library, Made in Punjab, Farzi Café	CDR/ FDR/ PBCL	Everstone	24	July 15	22.0%
4	Barbeque Nation Hospitality	Barbeque Nation	CDR	CX Partners	107	June 15	NA
5	BTB Marketing	The Beer Café	Café	Grantie Hill, Matfield, Harsh Mariwala	30	December 14	12.8%
6	Devyani International	KFC, Pizza Hut and Costa Coffee	QSR/ CDR/ Café	Temasek	500	October 14	15.0%
7	Blue Foods	Spaghetti Kitchen, Copper Chimney, CBTL	QSR/ CDR/ FDR/ Café	Verlinvest, Everstone	7	May 14	NA
8	Guha Roy Food Joint and Hostel	Only Alibaba and Baked & Fried	QSR	SEAF	15	March 14	28.0%
9	Faasos	Faasos	QSR	Sequoia Capital India	20	December 13	12.8%
10	Burger King India	Burger King India	QSR	Everstone	NA	October 13	NA
11	Barbeque Nation Hospitality	Barbeque Nation	CDR	CX Partners	110	March 13	29.7%

TABLE 5.4 *(Continued)*

S. No.	Target	Key brand(s)	Format(s)	Investor(s)	Amount (INR Crore)	Date	Stake
	Buyout						
1	deGustibus Hospitality	Indigo, Indigo Deli	CDR/ FDR	India Value Fund	190	July 15	NA
2	Sapphire Foods	KFC and Pizza Hut	QSR/ CDR	Samara, GS Investments, CX Partners, Others	205	July 15	NA
3	Moshe's Fine Foods	Moshe's	Café / CDR	New Silk Route	NA	September 13	NA
	PIPE*						
1	Speciality Restaurants	Mainland China, Oh Calcutta, etc.	CDR	SAIF	14	September 14	2.3%
2	Speciality Restaurants	Mainland China, Oh Calcutta, etc.	CDR	SAIF	35	January 14/ December 13	6.0%

Source: NRAI Report, 2016

5.2.3.1 M&A ACTIVITY AND EXIT DEALS IN THE FRONT-END SPACE

Both merger and acquisition activity and investor exit from businesses are signs of maturity in the food and beverage service industry. For nearly a decade, the segment received significant funding as shared earlier in this chapter and charted a strong growth curve both in terms of the number of outlets overall and the number of new entrants that joined the ranks. As the market matured, it was only natural for some level of consolidation to take place. The table below documents some key M&A transactions that took place and also some of the exits that took place as investors decided to take out their investments for varied reasons.

TABLE 5.5 M&A Activity in the F&B Front-end.

S. No.	Target	Key brand(s)	Format(s)	Investor (s)	Amount (INR Crore)	Date	Stake
1	Nilgai Foods	Pico Express	QSR	R&A Foods	30	July 15	100.0%
2	Hyderabad House	Hyderabad House	QSR	Food Krafters and Services	NA	May 15	NA
3	Global Franchise Architects	Pizza Corner	CDR	Avan Projects	NA	November 14	80.0%
4	Barista Coffee Company	Barista	Café	Carnation Hospitality	NA	August 14	100.0%
5	Love Sugar and Dough	Love Sugar and Dough	Café	Speciality Restaurants	1	May 14	51.0%
6	Nalas Aappakadai	Nalas Aappakadai	CDR	Epicurean Events	NA	May 14	51.0%
7	Om Pizza & Eats	Papa John's	QSR	Avan Projects	NA	December 13	80.0%
8	Idli Dosa	Idli Dosa	QSR	Cookie Man	NA	August 13	NA

Source: NRAI Report, 2016

5.2.4 F&B SERVICE INDUSTRY BACK-END RELATED INVESTMENTS AND MORE

Without that food and beverage, back-end kicking in parallel at a pace with which the front-end has grown over the decade the growth story would not have reached where it has. The back-end infrastructure for the food and beverage service industry in India was relatively weak in terms of the cold chain, food processing units, and even the storage aspects that were much needed to support the front-end delivery at a standard that global brands would expect. This has been an opportunity for local investors to latch on to as the demand side presented itself.

The tables below provide data cited in NRAI Report, 2016, it is apparent that substantial private equity investments found their way toward expansion and creation of the back-end infrastructure, especially,

TABLE 5.6 Key Exits/Liquidation Events in the Front-end Space.

S. No.	Target	Key brand (s)	Format (s)	Exiting Shareholder (s)	Exit type	Investor(s)	Amount (INR Crore)	Date	Stake
1	Coffee Day Enterprises	CCD	Café	KKR, NSR, Standard Charted PE, Others	IPO	IPO	1150	October 15	17.0%
2	Devyani International	KFC, Pizza Hut and Costa Coffee	QSR/ CDR/ Café	ICICI Venture	PE Secondary	Temasek	500	October 14	15.0%
3	Specialty Restaurants	Mainland China/Oh Calcutta	QSR	SAIF partners, Glix Securities	IPO	IPO	176	2012	25.0%
4	Jubilant Foodworks	Domino's	QSR	JP Morgan and India Private Equity Fund	IPO/ OFS	IPO	329	2010	35.6%

Source: NRAI Report, 2016

the food/beverage processing units and logistics elements like the cold chain. The data shared below speak for itself, as the industry continues to grow and now expand to tier 2/3 cities and semi-urban India and eventually the rural hinterland as well over a period of time, the back-end will take a cue from the front-end growth and continue to demand more private investment. While it may look like a large private investment play, the role of the government with respect to investments in the F&B Industry back-end is critical and vast. Policy level support, tax breaks, and multiple other SOPs have over the years encouraged investments in the back-end related businesses. governments at state as well as center are supportive and recognize the huge overall employment impact the food and beverage services businesses can make and thus either through tourism policy or initiatives for SMEs or as part of the larger employment generation agenda through support for entrepreneurship. at least on paper, there is support toward setting up both front-end and back-end businesses.

TABLE 5.7 Key Private Equity Investment in the Back-end F&B Space.

S. No.	Target	Segment(s)	Investor(s)	Amount (INR Crore)	Date	Stake (%)
Food ingredient/processed foods early stage						
1	Raw Pressery	Beverages (fruit juices)	Sequoia Capital India	31	December 15	30.8%
2	Veeba Food	Processed foods (Sauces)	Saama Capital, Others	25	July 15	14.5%
3	Arohan Foods	Food processing (Meat and Pork)	Omnivore Partners	3	April 15	33.4%
Growth stage						
1	Cremica	Processed foods (Sauces)	Rabo Equity	100	June 16	NA
2	Bira 91	Beverages (Beer)	Sequoica Capital India	40	January 16	20.0%
3	Maiyas Foods	Processed Foods	Peepul Capital, Ascent Capital	200	May 16	NA
4	Cremica	Processed foods (Sauces)	CX Partners	390	August 15	NA

TABLE 5.7 *(Continued)*

5	Manna Foods	Processed foods (Corn Flakes, Pickle)	Fulcrum Ventrue India, Others	30	May 15	34.0%
6	Bakers Circle India	Food Ingredients (Bakery Products)	Gaja Capital	30	February 15	18.0%
7	Himadri Foods	Food Ingredients (Masala)	SEAF	20	March 14	27.6%
8	Citrus Processing	Fruit juice manufacturing company	Cargill Ventures, Others	220	July 13	70.5%
9	Capricorn Food Products	Processed foods (Pulps/ Purees, concentrates, etc.)	Quadria India	60	April 13	34.5%
Buyout						
1	VKL Seasoning	Food ingredients (Spices)	India Value Fund	NA	January 13	NA
	Logistics/ Cold Chain Growth					
1	GatiKausar	Logistics services (cold chain)	Mandala Capital Fund	30	October 14	30.0%
2	Snowman Logistics	Logistics services (cold chain)	Norwest	60	June 13	14.3%
Pre-IPO (Initial Public Offering)						
1	Snowman Logistics	Logistics Services (cold chain)	Faering Capital	16	September 14	NA

S. No.	Target	Segment(s)	Investor(s)	Amount (INR Crore)	Date	Stake (%)
Dairy early						
1	Sahaj Agro Farm	Milk, organic staples	SIDBI VC	16	January 15	NA
2	Shikhar Dairy	Milk products	SIDBI VC	2	November 13	NA
Growth						
1	Parag Milk	Milk, milk powder, ghee, paneer, cheese	IDFC PE	60	September 15	NA
2	Milk Mantra	Milk, curd, paneer	Fidelity Growth Partners, Aavishkaar	80	June 14	61.6%
3	Prabhat Dairy	Milk, milk powder, curd, ghee	Proparco	188	June 13	14.2%
4	Parag Milk	Milk, milk powder, ghee, paneer, cheese	IFC	93	May 13	NA
PIPE						
1	DFM Foods	Processed Foods	WestBridge	345	April 16	26.0%
2	Hatsun Agro	Milk, milk powder, curd, ghee, butter, ice cream	MCap Fund	NA	November 14	NA
3	Hatsun Agro	Milk, milk powder, curd, ghee, butter, ice cream	WestBridge	35	June 14	1.2%
Pre-IPO						
1	Prabhat Dairy	Milk, milk power, curd, ghee	TVS Capital	85	September 15	7.6%
Others early						
1	The Fish Chain	Seafood	P39 Capital	NA	August 15	NA
2	Natureland Organic Foods	Organic foods (pulses, cereals, juices)	SIDBI VC	15	March 15	30.0%

Growth						
1	Akay Flavours and Aromatics	Nuts & spices, Aromatic extracts	ADB	37	November 14	16.7%

Abbreviations: NA, not available; PIPE, private investment in public equity; pre-IPO, pre-initial public offering.
Source: NRAI Report, 2016

5.2.4.1 M&A ACTIVITY AND EXIT DEALS IN THE BACK-END SPACE

Both merger and acquisition activity in back-end related industries has been active in recent years. The number of mergers has been scarce, but there is a clear indication of investor activity through private equity and IPO (initial public offering) route. This is indicative of a phase of consolidation as businesses map their investments to growth pathways and profitability objectives.

TABLE 5.8 Key M&A Deals in the Back-end Space.

S. No.	Target	Segment(s)	Investor(s)	Amount (INR Crore)	Date	Stake (%)
1	Creamline Dairy	Milk, milk product, ghee, ice cream	Godrej Agrovet	150	December 15	25.0%
2	Tirumala Milk Products	Milk, milk product, ghee, paneer, cheese	Le Groupe Lactalis	1750	January 14	100.0%

Source: NRAI Report, 2016

5.2.5 SIGNIFICANT INVESTMENT IN THE FOOD SERVICE SPACE BY FUNDS

There are many success stories in the organized food and beverage service formats that owe its success to strong and sustained investments by funds (domestic and international). Table 5.10 provides a list of some prominent companies/brands that have made a mark in the Indian market in their respective segments. Chaayos, Mamagoto, and many others have made it because the investors kept the faith in the brand and the leadership teams,

TABLE 5.9 Key Exit/Liquidation Events in the Back-end Space.

S. No.	Target	Segment(s)	Exiting shareholder (s)	Exit type	Investor (s)	Amount (INR Crore)	Date	Stake
1	Parag Milk Foods	Milk, milk powder, curd, ghee	IBEF, IDFC, Promoters	IPO	IPO	751	May 16	40.7%
2	Prabhat Dairy	Milk, milk powder, curd, ghee	Raboban and Proparco, others	IPO	IPO	302	September 15	26.9%
3	Cremica	Processed foods (sauces)	Motilal Oswal PE/Bector family	PE Secondary	CX Partners	390	August 15	NA
4	Snowman Logistics	Logistics services (cold chain)	Mitsubishi Corp, IFC, Norwest Venture, Others	IPO	IPO	197	August 14	25.2%
5	Milk Mantra	Milk, curd, paneer	Aavishkaar, Angel investors	PE Secondary	Fidelity Growth Partners	80	June 14	61.6%
6	Tirumala Milk Products	Milk, milk powder, ghee, etc.	Carlyte, Promoters	Strategic	LE Groupe Lactalis	1,750	January 14	100.0%
7	Prabhat Dairy	Milk, milk powder, curd, ghee	ICICI Venture	PE Secondary	Proparco	188	June 13	14.2%

Source: NRAI Report, 2016

and were willing to sustain losses to achieve growth and volume in a market that is quite competitive. As listed in Table 5.11, the advantages of fund-based investment are that the start-up is not starved of funding prematurely. the funds provide guidance on the exit strategy and have a team of professionals to support the start-up founders. Though it is early days, it does look like that some of the brands are here to stay, carving out market share in a growing market.

TABLE 5.10 Prominent Investment by Funds in the F&B Service Space.

Front-end	Back-end	Foodtech
Azure Hospitality (Mamagoto) Massive Restaurants (Masala Library, Made in Punjab) Degustibus Hospitality (Indigo, Indigo Dell) Dodsal Group Chai Point The Beer Café Devyani International SagarRatna	Backers Circle Cremica VKL Seasonings Prabhat Dairy Snowman Logistics	Faasos Swiggy Zomato Yumist Holachef Freshmenu

Source: NRAI Report, 2016

TABLE 5.11 Considerations and Benefits of Private Capital Investments.

Key investment considerations	Key benefits of private capital
• Underlying strengths of a business • Promoter quality • Realistic exit plan • Fair deal terms	• Long-term capital • Robust corporate governance practices • Higher visibility among all stakeholders • Strategic and financial advice • Strategic alliances leveraging the investor's network of contact • A partial or total exit strategy

Source: NRAI Report, 2016

5.2.6 INTERNATIONAL OPERATING MODELS OF SELECTED INDIAN BRANDS

When it comes to understanding how various possible business models have played out in the Indian F&B service space in terms of ***international***

expansion, the two operating models that seem to be ruling the roost are "franchisee" and "company owned". Given the interest in Indian food and the sheer size of the Indian Diaspora, it makes sense for Indian brands to look at international expansion as a growth strategy in addition to the domestic growth. Table 5.12 lists some prominent examples, the data from NRAI Report, 2016 may not be valid currently, but it makes the point nonetheless.

TABLE 5.12 Prominent Indian F&B Service Brands and their International Operating Models.

Brand	First international outlet	Total number of international outlets	International presence	Operating model
Chor Bizarre India's Restaurant NEW DELHI • LONDON • LAVASA	1997	1	London	Company owned
Hotel Saravana Bhavan	2000	56	USA, Canada, UK, Australia, Singapore, Thailand, France, Kenya, South Africa, Germany Netherlands	Franchise owned
Bikanervala	2006	6	London, Dubai, Nepal, Singapore, New Zealand	Franchise owned
Moti Mahal	2010	8	Oman, Tanzaniam Saudi Arabia	Franchise owned
Oh! Calcutta	2013	4	London, Dhaka, Singapore, Dubai	Franchise owned company operated
Mainland China	2013	4	London, Dhaka, Singapore, Dubai	Franchise owned company operated

TABLE 5.12 *Continued*

Brand	First international outlet	Total number of international outlets	International presence	Operating model
Café Coffee Day	2013	16	Austria, Czech Republic, Malaysia	Franchise model
Barbeque Nation	2014	4	Dubai, Australia, New Zealand	Company owned
Punjab Grill Gourmet Fine Dining	2014	3	Singapore, Abu Dhabi, Thailand	Franchise owned
Kailash Parbat Chaats | Dining | Banquets	2014	4	Dubai, Singapore, London, New York	Franchise model
Zambar Coastal Kitchen	2015	1	UAE	Franchise owned
Indian Accent	2015	1	New York	Company owned
Farzi Café Modern Spice Bistro	2016	1	Dubai	Franchise model

Source: NRAI Report, 2016

5.3 HOW IS TECHNOLOGY IMPACTING THE FOOD AND BEVERAGE SERVICE INDUSTRY IN INDIA?

Technology has been at the forefront of the expansion of the food and beverage sector in India, driving many trends, supporting innovation, and facilitating the evolution of many formats of the organized segment of the business. As has been the case with organized retail in general, the typical brick and mortar formats of food and beverage services wherein customers would be either serviced at the retail front-end or facilitated through pick up, delivery or drive-through options, technology-driven innovation has created more complex operations. some of the popular ones are listed below:

(a) Virtual restaurants—No retail front-end in the brick and mortar form, online presence, menus, all relevant information for customers, mobile-enabled apps, customers can see menus, reviews, order and pay online and get food delivered. Only investment in production back-end, mostly at a fraction of cost as location becomes irrelevant.
(b) B2C food delivery aggregator—Platforms that connect multiple food and beverage service operations and customers digitally, and facilitate all aspects of a "virtual restaurant" for participating food and beverage businesses. This enables businesses to leverage the advantages and reach of the digital platform to grow revenues through these aggregators.
(c) B2B food delivery aggregator—They perform the same task as the B2C aggregators, but do that for businesses alone. eventually, the B2C connection is also made.
(d) Restaurant search, reservations, recipe sharing, and food ordering apps.
(e) Ready-to-eat/Ready-to-cook food delivery.
(f) Train food delivery.

None of the above listed would find play without the "tech" component. There are 250 plus foodtech start-ups that are driving these innovative models. The segment is instrumental in ensuring a strong growth in online food ordering, which by modest estimates is growing at 30% or more. Foodtech businesses have their own challenges in the form of heavy discounting and high customer acquisition costs. Their success is dependent on ensuring that acquired customers remain loyal, and become repeat customers justifying the costs. While these businesses have been growing rapidly, it's largely cash burn funded by investors. They need to achieve a break even and eventually start making money. Given the low margins, without scale this may not happen, putting a question mark on their very survival.

The change on the ground has been rapid, gone are the days when one needed to refer to restaurant guides to find out about them or refer to the local yellow pages. It's all tech-driven and the customers love it. Online food delivery, reservation-based bookings, etc. are part of this socio-cultural shift. The change can also be seen in terms of cuisine preferences as more and more urban, tech-savvy customers are willing to explore more cuisine options rather than good old Indian food. It is estimated that nearly 60% of dinning incidences at restaurants would be reservation based by 2021 as against the current 30%. The industry is warming up to these tech-driven trends and sees this as an opportunity to redefine their product offerings and save costs.

TABLE 5.13 Key Food Discovery Players.

Company	Inception	Player type	Operations	Funding (till April 2016)
Zomato	2008	Global	Pan India	USD 224 million
Burrp	2006	Indian	Pan India	USD 2 million
Trip Advisor	2000	Global	Pan India	Not available

Source: NRAI Report, 2016

TABLE 5.14 Key Players in Table Reservation.

Company	Inception	Player type	Operations	Funding
Eazydiner	2014	Indian	Four cities	USD 4 million
Dine out	2012	Indian	Eight cities	Not available
Eat2eat	2000	Global	Eight cities	USD million

Source: NRAI Report, 2016

TABLE 5.15 Key Players in Online Delivery Segment.

Company	Inception	Player type	Operations	Funding
Swiggy	2014	Indian	Eight Cities	USD 60 million
Zomato	2008	Global	Pan India	USD 224 million

Source: NRAI Report, 2016

5.3.1 INVESTMENTS IN THE FOODTECH SPACE

India is now recognized as start-up capital of sorts, with more than 10,000 start-ups across multiple business categories. The food and beverage service sector in India has seen investments in a range of start-ups covering multiple formats for both front-end and back-end related businesses or operations as discussed in the chapter earlier. Foodtech investments have also gained interest as the role of technology in driving the food and beverage businesses has become integral. Foodtech space that used to be interpreted as synonymous with food processing, preservation, or development is more about tech platforms that are driving "food discovery or restaurant search", "table reservations", and "online delivery".

A review of the data shared below will leave no doubt in one's mind about the scope and size investments made in the foodtech space.

TABLE 5.16 Prominent Foodtech Deals.

S. No.	Target	Sector(s)	Investor(s)	Amount (INR Crore)
1	Swiggy	B2C food delivery aggregator	Norwest, Accel India, DST Global	47
2	Petoo	Virtual restaurant	Axilor Ventures	7
3	Innerchef	Virtual restaurant	Endiya Partners	NA
4	FreshMenu	Virtual restaurant	Lightspeed Ventures, Zodius Capital	110
5	Innerchef	Ready-to-eat and ready-to-cook food delivery	Endiya Partners	NA
6	Swiggy	B2C food delivery aggregator	RB investments, Harmony Partners, SAIF, Norwest, Others	235
7	Eatfresh	Virtual restaurant	Kalaari Capital	NA
8	Yumist	Virtual restaurant	Unilazer Ventures, Orios VP, Others	13
9	Faasos	Virtual restaurant	Sequoia Capital India, Ru-Net Holdings, Lightbox	200

TABLE 5.16 *(Continued)*

S. No.	Target	Sector(s)	Investor(s)	Amount (INR Crore)
10	Chefs Basket	Packaged foods	SA IF, Others	40
11	Grab	B2B food delivery aggregator	Sixth Sense Ventures	NA
12	Fresh	B2B food delivery aggregator	Kae Capital, Others	7
13	Roadrunnr	B2B food delivery aggregator	Sequoia Capital India, Nexus Venture Partners, Apoletto	65
14	Opinio	B2B food delivery aggregator	AccelIndia, Sands Capital, Others	45
15	TinyOwl	B2B food delivery aggregator	Sequoia Capital India, Matrix Partners India	50
16	Innerchef	Ready-to-eat and ready-to-cook food delivery	TA Ventures, M&S Partners, Others	11
17	Travelkhana	Train food delivery	Astarc Ventures	NA
18	Grab	B2B food delivery aggregator	Zomato Media	7
19	Shadowfax	B2B food delivery aggregator	Eight Roads Ventures	50
20	Zomato	Restaurant search portal and B2C food ordering & delivery aggregator	Temasek, Vy Capital	388
21	EazyDiner	Restaurants reservation portal	Saama Capital, Others	20
22	BetterButter	Recipe sharing platform	M&S Partners, Others	2
23	Opinio	B2B food delivery aggregator	AccelIndia	6
24	Holachef	Virtual restaurant	Kalaari Capital, India Quotient, Ratan Tata	20

TABLE 5.16 *(Continued)*

S. No.	Target	Sector(s)	Investor(s)	Amount (INR Crore)
25	Roadrunnr	B2B food delivery aggregator	Sequoia Capital India, Nexus Venture Partners, Blume Ven	63
26	Swiggy	B2C food delivery aggregator	SAIF, Norwest, Accl India, DGST Global	102
27	Roadrunnr	B2B food delivery aggregator	Nexus Venture Partners, Blume Ventures	5
28	Grab	B2B food delivery aggregator	Oliphans Capital, others	6
29	Opinio	B2B food delivery aggregator	Accellndia, Others	2
30	Zomato	Restaurant search portal and B2C food ordering & delivery aggregator	Sequoia Capital India, Vy Capital, Info Edge, Others	311
31	Delightfoods	Food ordering	RB Investments	NA
32	Holachef	Virtual restaurant	India Quotient	2
33	Frsh	B2C food delivery aggregator	Kae Capital, India Quotient	3
34	Yumist	Virtual restaurant	Orios VP	7
35	SpoonJoy	Food delivery	SAIF	6
36	Swiggy	B2C food delivery aggregator	SAIF, Accel India	12
37	Faasos	Virtual restaurant	Sequoia Capital India, Lightbox	96
38	TinyOwl	B2C food delivery aggregator	Sequoia Capital, Nexus Venture Partners, Matrix Partners	111
39	TinyOwl	B2C food delivery aggregator	Nexus Venture Partners, Sequoia Capital India	13
40	Box8	Food ordering	Mayfield	14
41	FreshMenu	Virtual restaurant	Lightspeed Ventures	31
42	Zomato	Restaurant search portal and B2C ordering & delivery aggregator	Sequoia Capital India, Vy Capital, Info Edge, Others	329

TABLE 5.16 *(Continued)*

S. No.	Target	Sector(s)	Investor(s)	Amount (INR Crore)
43	WowTables	Restaurants reservation portal	Karan Bhagat, PrabhatAwasthi, and Atul Kumar	2
44	Frsh	Food ordering	India Quotient	0.3
45	TinyOwl	B2C food delivery aggregator	Sequoia Capital India, Nexus Venture Partners, Others	16
46	Justeat	B2C food delivery aggregator	Forum Synergies, Others	10
47	Zomato	Restaurant search portal	Sequoia Capital India, Others	228

Source: NRAI Report, 2016

5.3.2 MARKETING DIGITALLY—THE WAY FORWARD FOR THE FOOD AND BEVERAGE SERVICE INDUSTRY

The role of technology can also be seen impacting how hospitality businesses are marketing themselves. The digital footprint on marketing spends of big and small food and beverage service establishments is being evidenced. It is not that traditional marketing media are no longer relevant. For mass reach and brand visibility, traditional media is still the way to go, but for more focused marketing efforts, digital marketing tools are emerging as value drivers. Digital media provide quick response time and direct customer feedback at a fraction of the cost incurred on traditional tools. Focused digital marketing campaigns are yielding great results for hospitality businesses. Social media, videos, emailers, and display ads are helping food service brands not only to communicate but also to interact with their customers. It is not just about creating a buzz for one's offerings; it's about engaging your targeted customer base. What Noone et al. (2011) discussed as an imperative for hotels in terms of engaging with the digital media is equally if not more relevant for the food and beverage service businesses at large.

In the Indian food and beverage services industry marketing spends in the range of 4–6% of the total revenues that have been the norm. This spend may be higher at about 8–10% for some operations based on the life stage the brand may be in. These spendings have been growing as the competition has grown, also what is noteworthy is the split of this spend between the traditional and digital formats. The larger and more established brands typically allocate about 30% of their marketing budget toward digital media, while the smaller players are seen to deploy 70% or more of their marketing budget toward digital marketing tools. Given the number of people reachable through smart devices, it is no longer an option for any hospitality business to not invest in digital media, irrespective of the dining format typology. The table below provides a perspective on the split of the digital marketing spend in percentage terms on various popular avenues by the various food and beverage service formats in India.

TABLE 5.17 Percentage Spends on Digital Media Avenues.

	Display marketing	Mobile Ads/SMS	Search and Email	Digital video	Third party	Social media
Café	10		10		50	30
QSR	10		20	50		20
Casual dining	9	19	51	9	2	9
Fine dining	11	2	26	12	6	43
PBCL	10			10	40	40
Frozen dessert/Ice cream	14			57	14	14

Source: NRAI Report, 2016

5.3.3 THE BIG ONLINE SHIFT AND ITS FUTURE PROSPECTS

For a nation with more than 60% of its population in the age group of 35 years or under, it is quite impossible to not recognize that the digitally savvy generations of Indians would best be connected through online media. Restaurant chains and food service operators/brands have all invested in interactive websites, search engine optimization, and search engine marketing that are important strategic considerations, as businesses are investing to ensure that they have favorable web page visibility as compared to competitors. Websites are largely "mobile compliant" and F&B businesses recognize the need to focus on their "online reputation" and are investing in "online reputation management" (ORM) by using professional services. Digital marketing is generating word of mouth, reviews, referrals and leads for businesses, and validating the investments made every step of the way. New methods of online payments like wallets and "near field communication" (NFC) have closed the loop in terms of ensuring that all aspects around search, identification, and purchase can be done effortlessly in a secure manner. Food blogs and review platforms offer another information layer that enables consumers to evaluate options in an informed way. these can be a double-edged sword is another discussion.

tables 5.18–5.30 shared below demonstrate that different dining formats have taken to different social media platforms as their social media platforms of choice. new options are emerging, and it will be important for businesses to keep evaluating the success or impact of their campaigns and analyze results with an objective to reach their target customers and communities at a minimal cost. Google offers great tools for digital marketers to analyze data, evaluate performance, and strategize their digital outreach and spends. Technology itself is evolving and changing rapidly with increased interest in "big data" and "artificial intelligence" (AI) in general. It would be safe to assume that technology will have a tremendous impact on hospitality businesses globally and India would be no different from that perspective.

TABLE 5.18 Fan Base for Cafes on Facebook.

	Fans	Growth
Café Coffee Day	48,37,815	1.4%
Starbucks	8,68,335	126.3%
Barista	2,30,613	−1.9%

Source: NRAI Report, 2016

TABLE 5.19 Follower Base for cafes on Twitter.

	Fans	Growth
Starbucks	55,786	462.7%
Café Coffee Day	50,563	61%
Barista	6,225	49.6%

Source: NRAI Report, 2016

TABLE 5.20 Fan Base for QSRs on Facebook.

	Fans	Growth
KFC	7,437,925	
Domino's	6,629,126	0.4%
Mc Donald's	1,381,682	10.9%

Source: NRAI Report, 2016

TABLE 5.21 Follower Base for QSRs on Twitter.

	Fans	Growth
Domino's	116,798	63.4 %
KFC	56,260	36.5.%
Mc Donald's	23,775	636.2%

Source: NRAI Report, 2016

TABLE 5.22 Fan Base for CDRs on Facebook.

	Fans	Growth
Pizza Hut	4,630,921	19.8%
Barbeque Nation	818,022	44.4%
Chili's	119,153	58.3%

Source: NRAI Report, 2016

TABLE 5.23 Fan Base for CDRs on Facebook.

	Fans	Growth
Pizza Hut	4,630,921	19.8%
Barbeque Nation	818,022	44.4%
Chili's	119,153	58.3%

Source: NRAI Report, 2016

TABLE 5.24 Follower Base for CDRs on Twitter.

	Fans	Growth
Rajdhani's	4498	26.3%
Barbeque Nation	2369	15.3.%
California Pizza kitchen	2369	14.2.%

Source: NRAI Report, 2016

TABLE 5.25 Fan Base for FDRs on Facebook.

	Fans	Growth
Masala Library	28,114	6.5%
Barbeque Nation	26,565	20.9%
Chili's	23,146	11.6%

Source: NRAI Report, 2016

TABLE 5.26 Follower Base for FDRs on Twitter.

	Fans	Growth
Indian accent	7411	2.7%
Olive	2305	1.8%
Masala Library	1354	2.1%

Source: NRAI Report, 2016

TABLE 5.27 Fan Base for PBCLs on Facebook.

	Fans	Growth
Rude Lounge	212,273	
Hard Rock Cafe	139, 136	3%
Indigo Live	132,025	11%

Source: NRAI Report, 2016

TABLE 5.28 Follower Base for PBCLs on Twitter.

	Fans	Growth
Blue Frog	273,056	106.6. %
Olive	24,158	22.8.%
Masala Library	4116	0%

Source: NRAI Report, 2016

TABLE 5.29 Fan Base for FD/ICs on Facebook.

	Fans	Growth
Baskin Robbins	883,678	0.5%
Cream Bell	254,314	109.4%
Natural Ice Cream	147,023	6.9%

Source: NRAI Report, 2016

TABLE 5.30 Follower Base for FD/Ics on Twitter.

	Fans	Growth
Baskin Robbins	12,206	49.2%
Natural Ice cream	4120	10.2%
Gelato Italiano	1881	29.37%

Source: NRAI Report, 2016

5.4 MANAGING REVENUES THROUGH REVENUE MANAGEMENT

From "bar exchanges" with screens enticing customers to enjoy the variable pricing to hiring revenue analysts doing what they do to get the food and beverage operators to offer the right price, at the right time to the right customer at the right place. Though RM has been around for decades, assisting the airlines and hotel industries to become competitive. the leaders in these businesses got many components right, and one of them was the RM capacity. The food and beverage service space in India is warming up to RM and it is clearly going to be a strategic differentiator for operators in a market that is growing on the back of serious investments but with weak profitability and massive cash burn. Singh and Munjal (2014) recognized that evolved revenue management practices would drive competitive advantage that is today more true for the F&B service space than ever as the industry grapples with rate transparency that technology has unleashed. What Cross (1998) had predicted for the airlines and hotel businesses way back is today ever relevant for the competing food and beverage businesses in India. Market domination will require businesses to use technology in a manner where data can be analyzed and utilized for effective pricing decisions that ensure profitability.

5.4.1 THE INDIAN STORY—HOW DOES IT LOOK?

If profitability was the only measure of success, then all are not right with the growth story of the organized Indian F&B service formats. The NRAI Report, 2016 cites financial data that is at best limited in its scope on the fiscal performance of some of the listed Indian companies. There is not much to show for in terms of profitability. While some have been able to maintain operating margins, it is only the cafe, QSR space that may be financially insulated at some level. The reason this information may mean little is the fact that a large chunk of F&B businesses that are not listed are ignored. absence of reliable financial performance data on these makes it impossible to comment on the financial health of this segment of the hospitality sector in India. Table 5.32 provides an international benchmark of sorts to put some additional perspective. clearly, the Indian story is yet to achieve both the scale and profitability that global players in the F&B service space can be expected to offer. The focus of the industry seems to be on "cash burn" to achieve scale and market penetration and eventually healthy profits.

As long as the "funds" are and remain interested, this should be fine. what the businesses will need to do is to ensure that they protect margins and build efficiencies so as to achieve acceptable margins in the short term.

TABLE 5.31 Listed Indian Companies.
All figures are in INR Crore except valuation multiples

Company	Outlet	Market Cap (May 31, 2016)	Enterprise value (May 31, 2016)	Financial data (FY 16)			Margins	
				Revenue	EBITDA	PAT	EBITDA	PAT
Jubliant Food Works (Domino's)								
Westlife Development (McD)								
Median							8.2%	2.3%
Coffee Day Enterprises	1607	5490	6896	2835	488	(8)	17.2%	NM
Speciality Restaurants	123	412	336	321	17	0	5.3%	0.1%
Median							11.3%	0.1%

Enterprise value = Market Cap ± Net Debt / (Surplus Cash)
Note: FY 16 (March 2016), EV = Market Capitalization +(-) Net Debt / (Surplus Cash), NM – Not meaningful
Source: NRAI Report, 2016

TABLE 5.32 Listed Global F&B companies.

All the figures are in USD million except valuation multiples

Company	Outlets	Market cap (May 31, 2016)	Enterprise value (May 31, 2016)	Financial data (FY 16)			Valuation multiples			Margins	
				Revenue	EBITDA	PAT	EV/Rev	EV/EBITDA	P/E	EBITDA	PAT
QSR											
McDonald's Corp.	36,467	1,07,990	1,27,830	25,360	9030	4820	5.0	14.2	22.4	35.6%	19.0%
Starbucks Corporation	23,921	81,420	83,000	20,160	4570	2540	4.1	18.2	32.1	22.7%	12.6%
Yum! Brands	42,725	34,120	38,010	13,100	2970	1320	2.9	12.8	25.8	22.7%	10.1%
Restaurant Brands International	4438	9900	17,900	4040	1630	162	4.4	11.0	61.0	40.3%	4.0%
Domino's Pizza	12,500	6390	8430	2250	435	192	3.7	19.4	33.3	19.3%	8.5%
Dunkin' Brands Group	19,471	4160	6380	815	409	117	7.8	15.6	35.6	50.2%	14.3%
The Wendy's Company	6482	2750	4940	1800	401	147	2.7	12.3	18.7	22.3%	8.2%
Papa John's International	4903	2430	2730	1630	181	80	1.7	15.0	30.6	11.1%	4.9%
Sonic Corp.	3153	1440	1920	619	171	70	3.1	11.2	20.6	27.7%	11.3%
Median							3.7	14.2	30.6	22.7%	10.1%
Fine dine											
Ruth's Hospitality Group	147	553	562	378	58	30	1.5	9.7	18.2	15.3%	8.0%
Del Frisco's Restaurant Group	60	372	373	338	44	16	1.1	8.5	23.2	13.0%	4.7%
Median							1.3	9.1	20.7	14.2%	6.4%

TABLE 5.32 *(Continued)*

Company	Outlets	Market cap (May 31, 2016)	Enterprise value (May 31, 2016)	Financial data (FY 16)			Valuation multiples			Margins	
				Revenue	EBITDA	PAT	EV/Rev	EV/EBITDA	P/E	EBITDA	PAT
CDR											
Daren Restaurant	1535	8580	8890	7020	908	337	1.3	9.8	25.4	12.9%	4.8%
Buffalo Wild Wings	1190	2770	2820	1880	287	99	1.5	9.8	28.0	15.2%	5.3%
Texas Roadhouse	491	3230	3230	1860	228	100	1.7	14.2	32.2	12.3%	5.4%
Brinker International	1647	2600	3730	3140	470	196	1.2	7.9	13.3	15.0%	6.2%
The Cheesecake Factory	201	2340	2420	2140	265	122	1.1	9.1	19.2	12.4%	5.7%
Bloomin' Brands	1502	2240	3540	4340	444	101	0.8	8.0	22.1	10.2%	2.3%
Median							1.2	9.5	23.8	12.6%	5.3%
Fast casual											
Chipotle Mexican Grill	2066	11,970	11,890	4250	665	327	2.8	17.9	36.7	15.6%	7.7%
Panera Bread Company	1997	5220	5520	2720	381	153	2.0	14.5	34.2	14.0%	5.6%
Fiesta Restaurant Group	365	648	732	700	97	38	1.0	7.6	17.3	13.8%	5.4%
Median							2.0	14.5	34.2	14.0%	5.6%

Enterprise Value = Market Cap ± Net Debt / (Surplus Cash)
Note: LTM (Last 12 months) as of March 2016
EV = Market Capitalization +(-) Net Debt / (Surplus Cash)
Source: NRAI Report, 2016

5.5 CONCLUSION

We live in interesting times as far as the food and beverage service industry is concerned in India. The investments poured in over the last decade or so backed by a promise unmitigated the growth and eventual profitability for investors and founders of the crowded start-up pile. As discussed in the chapter, the business models were built around sources of capital–private equity, venture capital, domestic/international funds, and to some extent investments made by Indian companies as they looked to diversify into another line of business. The revenue growth aside, the profitability seems to be a mirage; nearly all start-ups in the space have been losing money and going through some serious cash burn. At some point, the funding taps are likely to run dry, leaving these operations to sink or swim based on where they are at in the business life cycle curve.

If a recent move by numerous city chapters of NRAI to logout from the various online ordering/discount platforms is any indication (The Economic Times, 2019), all is not well the industry. This is even more critical as we are witnessing the lowest inflation rates over a decade and labor costs have not gone up at all. The inability of organized F&B service formats (chain as well as standalone) to remain financially viable without external funding must be resolved. These businesses will need to take a look at their business models, operating costs, and their expansion plans to ensure that they achieve fiscal health in tandem with market share, and not one at the cost of the other.

The financial performance in terms of net profit margin, market capitalization, and net worth of global F&B service companies as listed in Table 5.32 is indicative of where the Indian F&B operators can go from here. It may be a long road, but clearly there is a huge underlying potential that needs to be unlocked.

KEYWORDS

- **F&B business models**
- **franchise**
- **food aggregators**
- **F&B digital marketing**

- **F&B revenue management**
- **foodtech**
- **private equity**
- **fund investment**

REFERENCES

62nd Annual Report (2017–2018) *Federation of Hotel & Restaurants Association of India.*

Annual Report (2016–2017) *Ministry of Tourism*, Government of India.

Annual Report (2017–2018) *Ministry of Tourism*, Government of India.

Bannerjee, R.; Munjal, S. Sustainable Restaurants - Current Status and Future Path. In *Managing Sustainability in the Hospitality and Tourism Industry*; Jauhari, V., Eds.; Springer: New York,2014, pp 387–430.

Bharadwaj, S. G.; Varadarajan, P. R.; Fahy, J. Sustainable Competitive Advantage in Service Industries: A Conceptual Model and Research Propositions. *J. Mark.* **1993**, *57*(4), 83–99.

Connolly D. J; Olsen M. D. An Environmental Assessment of How Technology is reshaping the Hospitality Industry. *Tourism Hospital. Res.* **2001,** *3*, 73–93.

Cross, R. *Revenue Management: Hard core tactics for market Domination.* New York: Broadway Books, 1998.

Cross, R. G.; Higbie, J. A.; Cross, D. Q. Revenue Management Renaissance: A Rebirth of Art and Science of Profitable Revenue Generation. *Cornell Hospital. Rep.* **2009**, *51*, 53–67.

Dev, C. Operating Environment and Strategy: The Profitable Connection. *Cornell Hotel Restaurant Admin. Quarter.* **1989,** *30*, 8–13.

Engel, J. F.; Blackwell, R. D. *Consumer Behavior*. Dryden Presee: New York, NY, 1982.

Euromonitor. *Full Service Restaurants in India.* [Online] 2015. http://www.euromonitor.com/.

FICCI Report. Unlocking the Potential in Food and Beverage Services Sector. [Online] 2015, http://ficci.in/spdocument/20589/F&B-Report.pdf (accessed Jun 12, 2018).

Hospitality Trend Report, Published by AF and Company Consulting IBEF.ORG, India Brand Equity Foundation, [Online] 2018, https://www.ibef.org/download/Tourism-and-Hospitality-January-2018.pdf (accessed Apr 20, 2018).

IBEF.ORG, India Brand Equity Foundation, [Online] 2017, https://www.ibef.org/download/Tourism-and-Hospitality-January-2017.pdf (accessed Apr 5, 2018).

IBEF.ORG, India Brand Equity Foundation, [Online] 2017, https://www.ibef.org/industry/retail-india.aspx (accessed Jan 4, 2019).

Indian Restaurant & Food Service Industry–Structure & Prospects, 2019 [Online]. Available from: http://www.careratings.com/upload/NewsFiles/Studies/Restaurants%20%20QSRs%20May%202019.pdf, (accessed Jul 10, 2019).

India Hospitality Industry Review, [Online] 2018, Published by HVS- Anarock, https://www.hvs.com/article/8488-hvs-anarock-india-hospitality-industry-review-2018 (accessed Apr 8, 2019).

Indian Hotel Industry Survey (2018), Published by Federation of Hotel and Restaurant Association of India in consultation with Hotelivate.

JLL, India Hospitality Review, [Online] 2017, http://www.jll.co.in/india/en-gb/Research/Ready-steady-go-India-hospitality-industry-review-2017.pdf (accessed May 10, 2018).

Latham, S. F.; Braun, M. R. Jilted? The manager's little book for keeping customers in a weak economy. *J. Business Strat.* **2010,** *4*.

Munjal, S.; Bhushan, S. The Indian Hospitality Industry: Dynamics and Future Trends; Apple Academic Press: New Jersey, USA, 2017.

NRAI, India Food Services Report (2016), Published by National Restaurant Association of India.

Noone, B. M.; McGuire, K. A.; Rohlf, K. V. Social Media Meets Hotel Revenue Management: Opportunities, Issues And Unanswered Questions. *J. Rev. Pric. Manage.* **2011,** *48*, 231–245.

Singh, A.; Munjal, S. Issues and Concerns in Technology Decisions: the Hotel

Operator's Perspective. Worldwide Hospital. Tourism Theme. **2012,** *4* (2), 174–184. DOI: https://doi.org/10.1108/17554211211217343.

Singh, A.; Munjal, S. In *Emerging Revenue Management Practices and its Challenges: Indian luxury Hotelier's Perspective*, Proceedings of the Global Competitiveness and Corporate Governance Imperatives in Emerging Economies, Ansal University Gurgaon, India, Feb 15–16, 2013; Batra, S., Vikas, S., Eds., 2013, pp.628–638.

The Economic Times, Restaurant brands log out off food aggregators. [Online] 2019. https://tech.economictimes.indiatimes.com/news/internet/300-prominent-restaurant-brands-log-out-off-food-aggregators/70684991he Economic Times, (accessed Aug 15, 2019).

CHAPTER 6

Traditional and Slow Food in India

ABSTRACT

The commercialization of agriculture in modern India is increasingly changing eating preferences amongst Indians; increasing nuclear families, growing cities, and disposable income are adding to the cause, more and more people are looking for convenience food, making it a preferred choice. But, this was never the case; India as a country was always known for its regional food, unlike Western countries. Indian recipes and eating habits were never documented well; therefore, with every city, every state the recipes, taste, and preference change. To understand the traditional Indian food in and its evolution, we will have to look at Indian food history and eating habits during the times and the influences of traders and invaders who approached India over the period of time. The food that Indians eat today has evolved over time, from ancient Indian civilizations dated back to 2500–1800 B.C. to modern India, the food has evolved in stages. The journey of traditional Indian food from Indus valley civilization to Vedic times followed by the contributions of ethnic communities like Buddhist, Jains is remarkable; later came the Mughals, Europeans, and finally the British to contribute to the traditional Indian food to develop it to what we today refer to as traditional Indian food (Prakash, 1961). The chapter is about the transformation of Indian cuisine, there are similarities in the eating habits of traditional Indian food and the ethos of the slow food movement that started in the 1980s in Italy. The study will at the past Indian tradition, culture, eating habits and will relate it with the philosophy of slow food. It will also look at the status of slow food in India and its future.

6.1 INTRODUCTION TO TRADITIONAL INDIAN FOOD

The Indus valley civilization grew around the river Indus, Mohenjo-daro and Harappa remained the two biggest cities during the times, wheat was the staple grain and the use of other food items like coconut, pomegranate, banana, and chickpea is also documented. Some of the cooking utensils used then resembled the modern utensils; small clay oven traced that resembled modern clay oven or tandoor, flat cooking plates of clay resembling modern *tava* indicated the cooking of flatbreads. The Aryans then did not change the food composition but agriculture and growing of rice, wheat, barley, and lentils grew during their times. They majorly settled around rivers which added the use of fish in their daily diets, melon and cotton also grew during Aryans and the documentation suggests that the Aryans were meat-eaters. With the growth of farming, the domestication of cattle like cow, buffalo, sheep, and goat also increased and became an integral part of Vedic food, use of milk especially cow milk and its by-products like curd, ghee, and buttermilk is seen in abundance, the use of goat and buffalo milk is also mentioned. The scripts from Yajurveda indicate the use of oilseeds, cereals, and pulses, especially *urad, masoor, moong, and matar*. Rigveda shows the use of mango, lotus stem, cucumber, and bottle gourd. The mention of honey in Rigveda shows that honey was the primary sweetener during the times; however, sugar cane is also mentioned, but no use of sugar is seen and during the Buddhist period honey got replaced with jaggery as the primary sweetener (Achaya, 1994). Buddhism and Jainism are believed to be a part of the Hindu religion; therefore, the eating habits of these communities resemble Hindu eating habits.

The development of cuisine was largely dependent on the local and seasonal availability, but there was another dimension to the traditional Indian food; Ayurveda, the science of life, where Ayur means life and Veda resembling knowledge. It is a traditional Indian system of using food as medicines. The philosophy of eating food as medicine is based on the balance in bodily systems. Ayurveda believes that every individual's physical being is derived from the combinations of three Doshas (life forces) *Vata*, *Pitta,* and *Kapha* and a disbalance on any of the three life forces would disturb the body. It is very important to maintain the natural balance between all three Doshas to be mentally, physically, and emotionally fit.

Cooking of food and its assimilation time plays an integral role in Ayurveda, the use of local ingredients, freshly available food, and easy to digest is considered essential in maintaining natural body balance. These rules are similar to the philosophy of slow food and slow food movement. Let us discuss a little deeper on the ancient Indian cooking system which will help us in understanding the historical perspective of slow food in India. Indian food culture is one of the oldest cultures in the world, it always believed the Vedas to be the main source of knowledge, In Indian scriptures there are four Vedas, Rigveda, Yajurveda, Samaveda, and Atharvaveda and the Ayurveda is believed to be a part of Atharvaveda. The fundamental principle of Ayurveda is to regularize absorption, assimilation, and elimination of food from the human body. The solid part of the body is made up of substances related to earth and the liquid part constitutes water, the empty spaces in body are occupied by air. Therefore, the key elements to maintain the balance in body are earth, water, and air (Morningstar, 1995).

The ideology of Ayurveda food categorizes food as *sattvic* (pure and freshly available), *rajasic* (fiery foods), and *tamasic* (spoiled food items). It encourages the eating of sattvic food, which is seasonally available, freshly plugged, and locally available. It provides a soothing effect on the body and mind. The rajasic foods are spicy foods that provide warmth to the body and as an after effect aggravate passion and aggression in the body. The third category of food is tamasic food, which includes meat, poultry, fish, alcohol, etc. This category of food disturbs the natural balance of earth, water, and air within the body and creates a state of confusion; therefore, it is advised to avoid tamasic food.

6.2 SLOW FOOD

The journey started from a small town of Italy, Bra, when a group of young people during their visit to the town realized the need for preserving local culture and traditions. This was the time during the 1980s when global political scenario was going through turmoil, but the group lead by Carlo Petrini fought against all odds and went on to create "ARCI" an acronym for Associazione Ricreativa Cultural Italiana. The main idea for the group to come together was to preserve and promote local culture, festivals, food, and wine. While the idea was already in place, the formation of an official organization to register it as a nationwide movement took place in 1986, which later on became a global movement. It gained momentum when McDonald's entered Italy. In fact, the term slow food was coined to oppose the entry of fast food in Italy. The philosophy of slow food is exactly the opposite of fast food is all aspects and looking at the size of the fast-food chain, it was difficult to convince people and spread the message across. However, the idea was not to compete with McDonald's or not let them expand, but primarily it was to defend regional traditions, good food, gastronomic pleasure, and slow pace of life (Slow Food, 2019).

In more than two decades of history, slow food has become a global mass movement, with the spread of more than 160 countries involving millions of people passionate about preserving biodiversity. The network has around 100,000 slow food members including farmers, chefs, fishermen, researchers, and anyone who is fond of good, clean, and fair food. The philosophy of slow food has three pillars, good, clean and fair, good quality, wholesome and flavorful food which is free of any chemicals, does not harm the environment, and fair for the producers with an accessible price for the consumers. Over the period of time the movement has evolved; different programs have been designed to work in different areas. For instance, the ark of taste aims to save endangered foods and defend gastronomic traditions, food, and taste education, understanding and educating the backward integration. Apart from such projects, there are a number of theme like climate change, family farming, bees, food waste, GMO, land grabbing, slow cheese, slow fish, slow meat, slow wine, slow cities. All these themes are indicative of the holistic approach that slow food intends to cater, the effort is to maintain proper balance in the ecosystem. Ambelopoulia and Ortolan are two of the classic examples where very little care of birds is required (Appadurai, 1996), as the availability of birds was in abundance, but with time birds are shrinking and the number for endangered species is increasing (Viegas, 2010). These are the symptoms of misbalance created in the ecosystem and much of it is due to human actions. The slow food movement prospers to resolve such issues, but it becomes difficult to execute plans across nationalities because of local political and geographical challenges (Elliott, 2004). In fact, the story of slow food started with *Thrush*, a type of game bird hunted, roasted, and served as a delicacy, but there is no segregation of capturing birds where the birds that are intended for consumption can be killed and endangered species are preserved.

The slow food movement and its founder Carlo Petrini are trying to give a new dimension to gastronomy; the idea is not to oppose the modern agricultural techniques, but evaluate them against the ethos of the organization and encourage or discourage them accordingly. The prime role of the organization is to preserve local culture and biodiversity that will lead to the betterment of the global environment (Schneider, 2008). The idea is not to oppose McDonald's but to create awareness about producing and eating healthy food. We need to know what is good, clean, and fair food, from where our food is coming (Petrini, 2007).

6.3 FOOD: MEDIUM OF PRESERVING TRADITIONS

In line with the slow food philosophy, the role of cooks becomes very important. Transferring recipes and its documentation is not only documenting food, it is the culture and tradition that is getting preserved in the form of food. Women who are involved in daily cooking or other household works are very closely linked with traditions and have raised the need to preserve them (Benn, 2012). With the advent of technology, global distances are shrinking and societies are getting multicultural, the interactions between societies provide traditional information and exchange knowledge, in the process food gets transferred and people get to know about other cultures (Flowers and Swan, 2012). But, according to Ritzer (2008), McDonaldization of the society states that eating habits are increasingly dominated by standardization and homogeneity. Repetition of recipes or dishes based on the seasonal availability is another factor that leads to the formation of the food tradition, the practices and norms become customs and recipes become traditional food. If food is so important and is closely linked to the formation of cultures, then the cooks, chefs, and connoisseurs play an integral role in the creation and preservation of regional and traditional food. In the modern society where convenience food is dominating society, it becomes increasingly important to teach and learn regional and seasonal food. The culinary professional need to spend more time in researching the lost recipes, young family members need to spend time with elderly people, so that the food culture gets transferred to the next generation. Warde (1997) argues that cultures are not fixed; most of the aspects change with time, in support Sutton (2001) believes that even in the most enriched cultures the methods of traditional cooking and customs are largely dependent on circumstances; therefore, the historical knowledge is a compilation of adoptions. They found adoption and improvisation are vital for any cuisine to be sustained for a longer period of time.

6.4 SLOW FOOD INDIA

Looking at the literature review so far, it is quite apparent that traditional Indian food has ethos similar to what slow food has to offer. The theory of good, clean, and fair food is exactly what the Indian literature has to offer, the use of regional and seasonal ingredients remain at the core. Following and preserving food culture and slow pace of life are the key features of both traditional Indian and slow food. But in recent years, Terra Madre Day (Mother Earth Day) has gained popularity globally. The attributes of slow food are increasingly becoming the main concern of the planet earth. Ever since the movement started, it has received support across sectors; farmers, chefs, and activists are getting involved in all possible ways. Picking up clues from International trends, this chapter is not far behind. Slow food has become the talk of the town, from Mumbai to Delhi to Shillong, five-star hotels to stand-alone restaurants, pop-up at homes, garden restaurants everyone is considering planning menu on slow food. Although in growing cities of India the concept of convenience food has gained popularity, but in smaller town even today people go to sabji mandi for fresh vegetables, they get their grains done from the mill, mustard oil is freshly extracted and used as the main cooking oil, and fresh vegetables are grown in the kitchen garden (Vishal, 2015).

In India, Meghalaya is leading the movement from the front and Sikkim is declared as the first and only state in India as an organic state, all the vegetables, fruit, herbs, etc. are completely free of any chemicals or pesticides. In March 2015, ten chefs from Meghalaya prepared a special menu for Chef Fabio Antonini, a leading slow food chef from Amsterdam. The menu was curated by the tribes of Meghalaya to create awareness and the need to preserve the regional food. Another such example is the special nice course menu by Chef Sujan Sarkar in Olive Bar and Kitchen. It is a special menu from his studio kitchen that caters to only 20 people at a time and all the raw materials are sourced from local farms. This not only attracts customers but also encourages local farmers to grow more of the regional ingredients. The increasing demand for domestic products has resulted in many domestic locally produced markets, from Societe Naturalle honey to 4S milk, from Tijara farms to DOFM (Delhi Organic Farmers Market), all these growing trends are the indicators of demand for organically produced seasonal and regional products. However, to use the local products regularly in menus, the quality and supply have to be consistent, but the main reason for local items not seen in the menu is the quality at par with the imported materials. For instance, Chef Manish Mahrotra's biggest challenge is the quality procurement domestically. He candidly asks, if I serve you "Yamuna Fish" in the menu, would you eat it (Vishal, 2015). In the entire menu, there are only three nonvegetarian dishes, consistency and quality of the supply is the main reason for not having more nonvegetarian items in the menu. Chef Manu Chandra from Olive Bar Bangalore believes that while we may not be able to produce best of the ingredients, but most of the best produce from India gets exported to other countries; we hardly get to see Hapus from Ratnagiri, Dusheri of Malihabad, chilies from Rajasthan or lime from Rangpur readily available in local markets or even the hyper marts in India. He adds chefs can visit local mandis for fresh produce, can work to streamline backward integration but only a handful of chefs can afford to do so. Some of the demand for Indian and foreign vegetables is met locally from the nearby farms which are nowadays started by young professionals or senior people working for the betterment of society, but because these farms are high-end farms and are not run by farmers so the prices are also quite high.

It is not only about the food that is being prepared in the restaurants, even the preservation staples and grains are an integral part of slow food. After the green revolution, the yield of most of the crops increased significantly. The idea was to meet the increasing demand of the country, but in the process, many of the traditional crops were lost. Traditionally if you look at, according to the National Bureau of Plant Genetic Resources, India was home to more than 167 varieties of crops, more than 320 varieties of wild edible crops, and many varieties of domesticated animals originated in India. There are more than 50,000 varieties of rice, around 1000 varieties of mangoes, and 5000 varieties of sorghum (Dutt, 2015). In Chhattisgarh, there are many varieties of rice with varied properties. They can grow at different water levels, high in protein, and has medicinal values. The best part is that these varieties are not developed in laboratories. They are a result of years of hard work from the farmers of the state. Even other states like Kerala, Madhya Pradesh, Odisha, and Tamil Nadu have a wide variety of cereals and grains. Realizing the need and potential of the slow food movement, Carlos Petrini along with more than 600 slow food leaders visited Shillong to celebrate the 2015 edition of slow food movement (Dutt, 2015). In the same year under the guidance of Joel Basumatri an active slow food activist and a member of the chef's alliance, the Slow Food Youth Network Nagaland was formed.

Kunal Chandra, a food and beverage professional and a graduate of slow food university, Italy feels that in India concept of slow food is in build and was always in our heritage, but to make it more acceptable in the modern society especially the organic produce, we need to make it affordable for the general public not only for the elite. He shares his experience in Italy. He was able to buy one carat of tomatoes from the local farmers market for one euro and used for making sauce for pasta and other dishes for almost three months, he adds, India needs to have similar things. But the trend for promoting and eating local food has already started. Earlier it used to be only in star category hotels or elite restaurants, but a lot of stand-alone restaurants have started offering healthy, locally produced items in their menu. Café Lota in Delhi is gaining popularity for serving regional Indian food with local ingredients, cuisines that are not very popular, like Uttarakhand, Himachal Pradesh, Uttar Pradesh, Assamese with notable dishes like kathal ke biryani, Kerala fish curry, and berry pulao (using local berries). Catering services like Carra in Delhi and The Table in Mumbai are known to serve food from local producers. Bohemian, a regional Bengali restaurant in Calcutta run by Chef Banerjee uses regional ingredients but offers fusion Indian food. Similarly, Chef Cardoz in Mumbai at The Bombay Canteen offers International cuisine using local Indian ingredients (Vishal, 2015).

Generally, food trends flow from five-star hotels to other food and beverage outlets; ITC Maurya Sheraton offers food based on seasonal and regional concepts, they have taken it to the next level. The dishes in the menu are color-coded, color representation is for "lost recipes," "heart-healthy," "local produce," and "forgotten grain," the color coding in fish offerings refer to endangered species or easily are readily available. The choice to opt healthy and contribute to the environment is on customers. It also educates and creates awareness amongst customers adds Chef Manisha Bhasin. The philosophy to serve healthy is largely driven by Chef Manjt Singh Gill, corporate chef, ITC hotels, who is also a supporter of slow food in India. It was his initiative to have whole wheat naan at Bhukara and using field fresh mustard in Mewati cuisine at Grand Bharat. In fact, the introduction of cuisine like Mewati in a five-star hotel is a big step in the promotion of local cuisine in line with the slow food philosophy. Chef Gill is also the founder president of the Indian Federation of Culinary Associations, his philosophy of food is similar to what Carlo Petrini projects through slow food. He believes that recipes are the last thing to learn in cooking and traditionally Indian food was always based on wellness and sustainability. Every state has a unique flavor that can only be achieved through local sourcing. Such dishes need no additional cream or butter and any flavorings to make it tasty. He adds that understanding seasonal food is important as it is derived from decades of experience of what grows in what season (Sidhu, 2018).

6.5 SLOW FOOD INDIA INITIATIVES

In India, the restaurant or the food business would not have come this far without the support of the local farmers, environmentalists, and volunteers. The recipes will only taste great if the raw material is of the highest quality; therefore, backward integration is of utmost importance. Along with the restaurants, the suppliers of organic food and sustainable products have also grown. It is debatable, if the restaurateurs created the need for the suppliers to grow or after gaining confidence in vendors, the restaurants started incorporating local and sustainable menus. Let us look at some of the initiatives that have helped grow the aspects of slow food in India.

6.6 NEED TO GO SLOW

We looked at gaining popularity of convenience food in India, but at the same time, people in urban cities are getting health conscious more than ever. According to Times of India (October, 2015), a leading newspaper of India showed that the amount of pesticide present in vegetables and fruits in Delhi NCR is way more than the permissible limit, the permitted MRL (maximum residue level) has increased two folds in the last seven years, In 2008–2009 only 183 out of 133,438 samples failed the MRL test; however, in 2014–2015, out of 20,618 samples 543 failed the MLR test, in six years the rise of pesticides in vegetables, fruits, meat, poultry, and spices has increased to 2.6% from 1.4%. Looking in details, the samples were picked from farmers local sabji mandis, retail outlets, supermarts, and nearby farms; 56% of samples collected were vegetables and they accounted the most, some of the vegetables which were high in MLR were coriander leaves, aubergine, tomato, ladyfinger, chili, cabbage, capsicum, and cauliflower. Similarly, samples were collected in other cities of India and the results were similar, some of the prime cities from where data was collected include Mumbai, Gurgaon, Port Blair, Hyderabad, and Jaipur. The samples were also collected from the farm gate and surprisingly the amount of MLR was found higher from the samples collected from mandis than the samples collected from the farm gate. The use of chemicals is not only at the time of growing but various chemicals are used in the ripening process of fruits. The farmers cannot wait till natural ripening so raw fruits are plucked and packed with chemicals to be supplied to retailers. Calcium carbide is generally used for the ripening process, the chemical has acetylene, a colorless gas and calcium cyanimide, typically used in fertilizers and both the chemicals are extremely harmful for human consumption (Mutha, 2015). It is becoming increasingly difficult for the consumers to get the kitchen greens free of chemicals. The impact of these chemicals is at multiple levels; retarded growth, heart diseases, stomach disorders, breathing issues, etc. The numbers are alarming, it is high time for all individuals across all levels to pay attention and put in efforts to not only consume responsibly but also empathize with the environment and follow sustainable practices. Let us look at some of the initiatives that have given hope for good quality of raw material, free from chemicals, and fair for all the stakeholders.

6.7 DELHI ORGANIC FARMERS MARKET (DOFM)

Sunday mornings at the poolside of The Park Hotel, New Delhi are the healthiest mornings of Delhi. It is not because they put extra air purifiers or artificial rain, but, because they host DOFM, a farmers market started by environment-friendly people that acts as a platform for farmers and consumers. It started off from Bikaner house with a few farmers from Bikaner and other parts of Rajasthan, before shifting to its current location The Park, a five-star hotel in the downtown of Delhi, very close to Jantar Mantar, a UNESCO World Heritage Site. Presently, farmers from across the country have joined the DOFM and sell their local produce, food items like elaichi banana, pahadi rajma, Kashmiri kahwa, etc. which were difficult to be sourced in Delhi are readily available. Having gourmet cheese, pure organic milk (4S), Kanji, millet, varieties of beans, fresh fruits, etc. are a very common sight (Fig. 6.1). Similarly DLF Club 5, on the Golf Course Road, Gurgaon hosts the Gurgaon farmers market. This place is very popular amongst the locals to but organic produce, especially jaggery, pulses, basmati rice, brown and red rice from Uttarakhand.

FIGURE 6.1 Visit to Delhi Organic Farmers Market.

6.8 NAVDANYA

Navdanya is an initiative by Dr. Vandana Shiva to protect and preserve biological and cultural diversity. Navdanya meaning nine seeds symbolizes the protection and preservation of traditions and cultures; it also means new seeds or the right to save seeds. Dr. Vandana believes that in today's time when biological destruction is so common, the seed savers are the true seed givers to the society. She adds, conserving seeds is like conserving biodiversity, conserving culture, conserving tradition, conservation knowledge, and conserving sustainability. Navdanya as an organization has achieved remarkably well in the past decade, it has trained over 900,000 farmers across country in seed and food sovereignty and sustainable agriculture. The network of Navdanya is spread across 22 states in India and has helped setup 122 community seed banks across the country. They believe in education youths and creating awareness amongst farmers; therefore, they have also stared Bija Vidyapeeth (seed or Earth

University) in Doon Valley, Dehradun. The initiative is a woman-centered movement for the protection of cultural and biodiversity (Fig. 6.2). It has also created awareness on the hazards of genetic engineering, food laws, globalization and climate change (Navdanya, 2019).

FIGURE 6.2 Navdanya, making life better.

6.9 HANS FOUNDATION

The foundation is a public charitable trust that supports nonprofit organizations across India. They primarily deal with health, education, disability, and livelihoods of underprivileged people across the country. Their significant amount of time goes into research, development, and training of the farmers (Fig. 6.3). They train farmers with techniques that are cost-effective, affordable, and at the same time sustainable (Fig. 6.4). The idea is to increase the yield with minimum investment so that the producers do not feel the need of adding any pesticides or chemicals during the farming process. The entire cycle from farm to retail is free of chemicals, only herbicides are used and all the green waste is converted into compost within the premises. The natural composting process ensures that all the minerals and important elements are preserved and are incorporated back in the soil. In the past ten years, they have helped around 7 million people in the country and have funded more than 300 nongovernment organizations.

With the vision to work toward creating an equitable society and with the aim of enhancing the quality of life, they played a major role in state-wide development in Uttarakhand, Nagaland, Manipur, Sikkim, Assam, Mizoram, and Arunachal Pradesh (The Hans Foundation, 2019).

FIGURE 6.3 Farmers training camp.

FIGURE 6.4 Bamboo replacing metal in farming.

6.10 CHAUPAL @ VEDATYA

Chaupal is an annual conference organized by Vedatya Institute, an institute for higher education in the domain of hospitality and culinary education (Fig. 6.5). This unique symposium acts as a platform for farmers, chefs, hospitality educators, students, nonprofit organizations, and any environmentalists who care for society and believes is adopting sustainable approach to life. It is typically a two-day event that offers an opportunity to display the organic produce; items like honey, rice, pulses, vegetables, fruits, milk, curd, homemade wines, and beer. Farmers from the surrounding areas get their produce and participate in the seminar; simultaneously, the industry chefs get a chance to interact with the growers which completes the food cycle of farm to fork.

FIGURES 6.5 Glimpses of Chaupal.

6.11 LEARNING FROM THE PEERS

Sikkim, an Indian state has become an inspiration after it was awarded the prestigious UN future policy gold award 2018. It has become the first-ever organic state in the globe; while accepting the award in the Food and Agriculture Organization of the United Nations headquarter in Rome, Chief Minister Pawan Kumar urged the world to build an organic world together. He added that the transformation has benefits at multiple levels, around 66,000 farmers are benefitted directly and it increases the consumption, expands the market, helps in rural development, and supports sustainable tourism (Fig. 6.6). It is important to train the youths and get the latest technology in practice without impacting the environment, with the adoption of the technology, Sikkim is able to source its organic produce to cities like Delhi and Bangalore. It is indeed an inspiration for all the other cities and states like Kerala, Himachal Pradesh, and Uttarakhand have already started following the organic footsteps.

FIGURE 6.6 Indian tribes celebrating.

6.12 REVIVING LOST INDIAN RECIPES

An integral part of the slow food movement is reviving and preserving lost recipes. Indian food has a long history from native dishes to historically adopted cuisines from around the world. In the Indian food journey, some of the recipes either got lost or with time, because of availability and socio environment factors some faded away (Fig. 6.7), but chefs are continuously researching and documenting the age-old recipes. Chef Ajay Chopra, Chef Kunal Kapur, Chef Manjit Gill, and Prof. Pushpesh Pant are some of the prominent names who have aggressively worked on lost Indian recipes, taking their efforts further, Chef Adiyta Bal is on a mission to unearth recipes from across India. He has got tremendous support from Daawat Basmati rice and was able to convey his work to common people through a TV show on an epic channel called "*Lost Indian Recipes.*" He believes that lost recipes are all about food that got diffused in the pages of history, with his current efforts he wants to rediscover the tradition that once was (Fig. 6.8). The episodes are very interesting and informative; they feature recipes from more than 13 Indian states, which are not found on cookbooks. Some of them evolved over time, but with his efforts, he wants to showcase and document the original Indian recipes.

FIGURE 6.7 Traditional Indian village.

FIGURE 6.8 Pickling vegetabes the traditional way.

6.13 CONCLUSION

Looking at the discussion, it is apparent that there are similarities between traditional Indian food and slow food from Italy. The way modern society is accepting slow food as a movement, it is becoming obvious that tradition, culture, and history are increasingly becoming innovation of present times. The documenting of food history and sharing of knowledge are important in intercultural exchanges. One of the main components of slow food is preserving traditions, traditional food, and lost recipes, but it is noticed that because of the industrial revolution, modern farming techniques, nuclear families, etc. the traditional cooking methods are disappearing. But one of the main problems with commercialization is that to increase the yield, the chemicals and fertilizers are used in abundance, in short-run it meets the need but in the longer run, the quality of soil deteriorates and chemicals are found even in the raw materials that are used in preparation of dishes.

As the slow food movement gains momentum all the stakeholders need to play their role, the chefs will remain at the center, the ones for creating awareness, initiating trends that will then be followed by masses, creating dishes from local ingredients that will appeal the taste buds of the larger audience, so if chefs showcase their skills, encourage local gastronomy and sustainable development, they can certainly make the difference. The suppliers, vendors, and retailors need to play their role, the idea is not to run a business, but to run a sustainable business with ethical practices, food handlers, farmers, growers, suppliers, and middlemen need to keep the environment, end-user in mind and ethical business has to be at top priority. Today's youth are going to be the future of the country, so there is a need to create awareness in schools, colleges, and all professional institutes, to make this a reality. Educators will play a crucial role, so environment, sustainability, ethical practices, and ethos of slow food need to build in the curriculum and to be taught at all levels. Michele Rumiz, a representative of slow food India, believes that the new generation of Indian youths is increasingly becoming aware and wants to make all required changes in the food system; therefore, the future of slow food in India is theirs.

We are living in a very fragmented society. We need to ask ourselves about social engagements and respond accordingly. The human race is facing environmental challenges globally, climate change, malnutrition, increase lifestyle disorders, lowering of the water table are at a record high, it seems that to a lot of global issues, we are the first and last generation who can think and act on before it gets too late. Historically if you look at, there was a balanced ecosystem, where all the species including human beings lived in peace and harmony, the environment was less polluted and human practices were sustainable. There was a time when the pace of life was slow, but with globalization there in no time, people are getting increasingly busy with their daily routines, resulting in disorders and frustrations at multiple levels. Is it time for us to learn from our forefathers? Do we need to slow down? Does the philosophy of good, clean, and fair food should be adopted globally and faster than anything? Is it time to develop understanding and create awareness about the importance of sustainable practices? Do we need to question ourselves that the efforts that we are taking are enough to preserve the environment? The list of questions to be answered looks endless, but before it takes over all of us, it is time for all of us to collectively act local and think global.

KEYWORDS

- **slow food India**
- **sustainability**
- **ethical practices**
- **lost recipes**
- **organic farming**
- **future of food**
- **food history**

REFERENCES

Achaya, K. T. *Indian Food: A Historical Companion*, Oxford University Press: Oxford, 1994.

Appadurai, A. *Modernity at Large: Cultural Dimensions of Globalization*. Minneapolis: University of Minnesota Press, 1996.

Benn, H. When Traditions Become Innovations and Innovations Become Traditions in Everyday Food Pedagogies. *Aust. J. Adult Learn.*, **2012,** *52* (3–9), 595–616.

Dutt, B. Why India Needs a Slow Food Movement, 2015. [Online] 2019. https://www.livemint.com/Opinion/fozrtmri5TfVVaYA2PFMpN/Why-India-needs-a-slow-food-movement.html (accessed Nov 1, 2018).

Flowers, R.; Swan, E. Eating the Asian Other? Pedagogies of Food Multiculturalism in Australia. *J. Multidiscip. Int. Stud.* **2012,** *9* (2), 1–30.

The Hans Foundation. [Online] 2019. http://www.thehansfoundation.org/ (accessed Nov 11, 2018).

Krishna Murthy, R. Cooking in Ancient India *Bhavan's J.* **1984,** 1st August, Bombay.

Morningstar, A. *Ayurvedic Cooking for Westerners: Familiar Western Food Prepared with Ayurvedic Principle*, Lotus Press Books: USA, 1995.

Mutha, S. Those Who Ripe Fruits with Chemicals are Terrorists- HC, 2015. [Online] 2018. https://timesofindia.indiatimes.com/city/hyderabad/Those-who-ripen-fruits-with-chemicals-are-terrorists-HC/articleshow/48549916.cms (accessed Nov 20, 2018).

Navdanya. [Online] 2019. http://www.navdanya.org/site/ (accessed Jan 1, 2019).

Petrini, C. *Slow Food Nation.* Rizzoli Ex Libris: New York, 2007.

Prakash, O. *Food and Drinks in Ancient India*, Munshi Ram Manohar Lal, Delhi. 1961.

Ritzer, G. *The McDonaldzation of Society 5*, Pine Forge Press: Thousand Oaks, 2008.

Schneider, S. Good, Clean, Fair: the Rhetoric of the Slow Food Movement. *Coll. English.* **2008,** *70* (4), 384–402.

Sidhu, L. India Turns to go Slow. [Online] 2018. https://www.tribuneindia.com/news/spectrum/india-s-turn-to-go-slow/690444.html (accessed Dec 22, 2018).

Slow Food International Statute. [Online] 2019. www.slowfood.com (assessed Feb 2, 2019).

Sutton, D. *Remembrance of Repasts: An Anthropology of Food and Memory*, Berg: Oxford & New York, 2001.

Times of India. [Online] 2019. https://timesofindia.indiatimes.com/india/Spurt-in-pesticide-laced-vegetables-across-India/articleshow/49220802.cms (accessed June 1, 2019).

Viegas, J. Millions of Songbirds Killed for Recipe. [Online] 2019. http://news.discovery.com/animals/ millions-of-songbirds-illegally-killed-for-pickled-poachedbird-recipe.html#mkcpgn=rssnws1 (accessed Feb 2, 2019).

Vishal, A. Slow Food: Indian Consumer Rediscovering a Taste for Local Ingredients, Over the Aspirational Imported. [Online] 2019. https://economictimes.indiatimes.com/magazines/panache/slow-food-indian-consumer-rediscovering-a-taste-for-local-ingredients-over-the-aspirational-imported/articleshow/46890595.cms?from=mdr (accessed Dec 1, 2018; Jan 5, 2019).

Warde, A. *Consumption, Food and Taste: Culinary Antinomies and Commodity Culture*, Sage: London, 1997.

CHAPTER 7

Cuisine Trends: Today and the Way Forward

ABSTRACT

Indian food preferences are changing more than ever. India is a multilingual country with people from different caste and creed. It is difficult to have one national cuisine; however, regional food has always dominated Indian states, but new food trends, like domestic and international are always visible across India. This chapter will look at the current cuisine trends and will explore the future of food in India. The Indian food journey stated with the regional and seasonal food. As the economy grew, international food chains started expanding and of all, fast food became most favorable. Numerous reasons could be discussed for the growth of fast food in India, but the focus of this chapter is on the popular food trends that are dominating Indian culinary business. It discusses about the transformation of Indian cuisines and highlights the prevailing trends. Toward the end, it gets concluded with the future of food in the Indian context.

7.1 INTRODUCTION

Indian food has a history of adopting eating habits from different societies, civilizations, travelers, and traders, but the tradition of having an international cuisine started in the early 1980s from The Taj Mahal Hotel, Mumbai, which was then followed by other five-star hotels. There was a time when international cuisine in India meant Italian or Chinese; rest of the cuisines that are now gaining popularity were unheard of. The culture of having a fine dining restaurant with international cuisine was restricted to five-star hotels only. This also meant that only selected customers could afford and such cuisines were far too away from masses. The idea was to differentiate the product and give customers the feeling of being in the country of food origin. The focus was more on taste, ambience, and décor and above all it was a status symbol, a way to segregate elite from masses. For the majority of Indians, the regional food with local ingredients and seasonal availability was still the basic norm. They had no other option but to eat healthy, and only for occasions or festivals, special meals were cooked which was more of a social ritual.

Ironically, the global food industry for ethical dining, which is estimated to grow more than 800 billion dollars has also identified two similar themes—food for health and food for social ritual as the main food trend that will dominate the food industry in coming future. The trend of consuming food as medicine looked new to the Western world, but in countries like Indian and China, it is an age-old tradition. With globalization and growing international concepts in India, it is more of creating awareness among Indian customers. Food is acting as a medicine for health issues that consumption of food has itself created. The awareness needs to be created between healthy and unhealthy foods and eating habits. There is increasing numbers of patients for lifestyle disorders, obesity, malnutrition, stress, anxiety, etc. All these factors have labeled food as to be ordered and not to be ordered. Some of the patrons with some basic understanding are still able to make a choice, but the majority of them are not confident enough about what to eat and what not. As a result, there is an enormous increase in professionals who could guide on diets and dietary requirements.

Over the period of time, these developments have made subtle changes in Indian cuisine and eating habits. Some of them are quite apparent, but some could not directly be related to food. The impact that it is having on cultural status of Indian food, which is outside the view of dieticians, food writers, and bloggers is not visible, but in the longer run, it will play a crucial role in the understanding of what Indian food is.

7.2 IS AUTHENTIC INDIAN FOOD AUTHENTIC?

Cuisine can be authentic in more than one ways; by definition, one could define a food that belongs to a particular ethnic group, that uses local ingredients without any interference or fusion of any other cuisine to be an authentic cuisine, anything else is inauthentic. Does it mean that the Indian cuisine that we eat today which has evolved over time is not authentic

Indian cuisine? Does customize recipes for commercial acceptance is not an authentic cuisine? If that is true, then, looking at historical evidences, it is really difficult to define what authentic Indian food is. India is by far one of the most diverse countries in the world; Indian food has adopted ingredients, recipes, cooking utensils, cooking methods, and dinning etiquettes from almost all the directions of the globe.

To understand the authentic Indian cuisine better, imagine Indian food without the use of potatoes and tomatoes. They never had Indian origin, so why an Indian meal is difficult to visualize without the use of these two vegetables? The most popular is rajmah chawal, where rajmah (kidney beans) is not of Indian origin. It is difficult to imagine koftas, gravies, kormas, and biryani without the use of cashew nuts. Sakar kand, bhutta, jimi kand, and papita were never grown in India, to an extent even the chili, pineapple, and chickoo were brought to India from South of America (Nandy, 2004). Tea, Pear, Cinnamon, Cherry, and Leechi were from China. Onion came from the Central Asia and Gobhi from Europe. Not only the fruits and vegetables but also a lot of spices that we use in Indian cuisine came from outside of India; ginger, garlic, asafetida, and turmeric are some of them.

A lot of dishes that have become an integral part of Indian cuisine today and globally getting popularized as Indian food are actually not of Indian origin. The famous Indian kebabs came from the Central Asia, but adopted Indian palate and got customized over the period of time; biryanis came from the West and Central Asia; generally, biryanis were nonvegetarians but vegetarian option got introduced in India. The other famous rice preparation pulao came from Persia where it is more popular as Pilaf, so does Jalebi. Even the world-famous South Indian idli came from Indonesia (Achaya, 1998). Although all these spices and dishes came from outside of India, their present use was developed in India, for instance use of asafetida and chilies in a pickle, and other spices in chutneys and curries. Potato might not have an Indian origin but the number of potato dishes that are prepared in India will easily overtake any other potato growing region in the world (Appadurai, 1998).

7.3 INDIAN SNACKS

Indian food is considered spicier than European foods and when it comes to snacking, they are even spicier than the usual meals. In the North of

India, samosa and pani puri dominate the list, but in recent past, some of the South Indian preparations are gaining popularity, like idli, vada, dosa, and uttapam. In fact, majority of the North Indians, when they go out for snacking prefer to eat South Indian snacks. One of the reasons is the healthy perception that is associated with steamed idlis and gluten-free dosas; however, in Punjab, stuffed paratha, choley bhatura, kulche, aloo tikki, chaat, and dahi bhalla are still popular. Also, because of the health issues and people getting health conscious, Gujrati snacks are making their way. Most of the Gujrati snacks use gram flour and are steamed and therefore gaining popularity (Fig. 7.1); dokhla, khangvi, and a variety of farsaans (Fig. 7.2) are some of them. At the same time, there is another group of people growing in numbers that do not relate themselves with the traditional Indian snacks, which reflects the concept of high and low cuisine, for example, eating burgers, sandwich, and pizza are popular in middle to a high class society and having the traditional Indian snacks is considered as low cuisine.

FIGURE 7.1 Gujrati snacks.

FIGURE 7.2 Farsaan.

7.4 FOOD TRENDS OF RECENT PAST

The year 2017 witnessed the rise of chefs and new trends. Most of the trends started in metropolitan cities, which then got copied in smaller cities and towns. Mumbai was all up for healthy, local, and regional cuisine with a lot of home-cooked foods and pop ups taking place. Vegetarian diets continue to dominate restaurants. This also resulted in decrease in packed and packaged foods, and the American cuisine was not the talk of the town. Instead, people were interested in exploring with the Asian cuisines. Burmese became the cuisine of the year for Mumbai. Use of gadgets increased in kitchens and some of the items which were believed to be elite like hummus, pita, shawarma, and tacos seemed moving toward smaller restaurants, and some of them even made their way to be seen as street food. In desserts, chocolate gained popularity and fruits moved from desserts on to the appetizers. There was a big rise in the gastropubs in Mumbai, where the food was served on platters and plates rather than silver service.

Delhi is typically known for the North Indian food, but the five-star hotels in Delhi NCR kept covering and innovating fusion cuisine. Earlier, it was the European food that was on the center stage, but this year, it was Indian food with a modern touch. The recipes used were traditional with local ingredients, but the presentation was much influenced from the modern art of French cuisine. The customers were getting more health conscious, and there was a rise in healthy eating. The chefs were experimenting with healthy cooking options. Therefore, the use of olive, mustard, peanut oil, and clarified butter (ghee) was evident in the menus. Along with the rise of modern Indian cuisine (Fig. 7.3), the pairing of food and wine was also seen increasing. Tacos are already a hit in Calcutta and past few years have seen a rise in Taco demand. There is awareness about the chemicals being used in food, and the use of plastic in the kitchen is getting avoided to the extent possible. Activated charcoal became the main ingredient and was used as a superfood. The use of technology like smoke guns and sous vide gained popularity, and chefs are increasingly becoming environmentally conscious. Therefore, the concept of having no dustbins was the general motto. The idea was to reduce the waste to a minimum and use the ingredients to the maximum. Bangalore always wins in entrepreneurial startups. So, even in the food business, food vans, small cafes, and small self-cooking businesses were noticed growing. The use of local ingredients even for the international menus was promoted. In 2017, Bangalore got the first brewery, which started a new trend of craft beer; the food served with it was more of platters from Indian, Mexican, Lebanese, and Chinese origin. Fresh vegetables, leafy greens, beans, and fruits dominated the menus (Mehta, 2018).

FIGURE 7.3 Traditional Indian food presented in modern style.

In general, food and beverage business in the last couple of years was exciting. Not only food trends dominated the Indian market, but there was also increasing awareness on health and healthy grains. Apart from specific cuisines and grains, there were certain trends that were seen growing in general (The Restaurant Times, 2018):

- **Whole grains:** In the last decade, refined grains dominated the menus but with increasing awareness, whole grains were seen in the menus. sugar got replaced with jaggery or stevia. Fresh fruits were preferred over refrigerated ones. it looked like the concept of whole grains is going to be the trend in coming years.
- **Baking:** with increase awareness about saturated and unsaturated fats, frying was reduced to the extent possible and baking became the cooking mode for most of the restaurants. in fact, some chefs removed the term fry or fried from the menu and some seen using stir fry.
- **Protein:** The is a sudden increase in the number of fitness centees across India and 2017–2018 looked like the era of fitness centers. this resulted in increasing the consumption of healthy protein like chicken and fish. red meats were avoided, but the inclination toward nonvegetarian protein was higher.
- **Barbecue:** Barbecue food became popular. rooftop restaurants opened up and the popularity of fixed menus was on all-time high.
- **Food festivals:** organizing food festivals started in 2017 and continued in 2018. festivals like grub fest, Sula feast, and Oktober-fest were hosted successfully and thousands of visitors were seen attending these festivals. a unique festival for all the street food vendors of Delhi NCR was organized to showcase the variety of street food that India has to offer along with the heritage food from Old Delhi.

7.5 CURRENT FOOD TRENDS

The previous years were full of new cuisines and food trends and 2018 was also looked the same. In fact, this year there was a significant increase in new concepts and trends compared to the last few years. Generally, when people become health conscious, they compromise on taste or quantity of food, but the kind of innovation that chef showcased, even without compromising on the taste, one can maintain their waistlines. The trend of home cooks started in 2018, there were pop ups or table of 6 or 4 being started by the home cooks. Their popularity showed the inclination toward light home cooked style food. Even in five-star hotels and restaurants, home cooks were invited to organize food festivals based on their regional specialty. Similarly, a lot of female cooks started home baking at home, taking online orders, and delivering at homes cake, bread, muffins, and brownie that were in demand. Regional Indian restaurants were opened in Indian metropolitan cities; cuisines that were not very popular in states like Assam, Orissa, Kashmir, Tamil, Uttaranchal, and Kerala, they were seen gaining popularity. With social media becoming as one of the main promotional tools, theme-based restaurants were opened. Themes like, Jungle safari, Lady Baga (Gaga), and while waiting gadgets to play games were provided to customers.

Unlike last year, where nonvegetarian protein was high on demand. This year, plant-based diets and plant proteins were on demand. Vegan diets started gaining popularity and people started to turn vegan. Plant-based recipes were seen replacing starchy foods; for example, zoodles and spaghetti made out of zucchini and zucchini ravioli continued to rule the menus (Fig. 7.4). Mushroom grew as a superfood with its natural presence of monosodium glutamate; it came as a healthy yet tasty ingredient in the menus. Tacos became the talk of the town with vegetarian innovation to it. The regular taco shells were replaced with lettuce, spinach, and kale. The regular taco shell was made gluten free and taco bell introduces chicken shell taco, which is yet to be replicated. Food in powdered form gained popularity. Turmeric, matcha, beetroot, and bell pepper powders were not only used as garnishing but also became a part of the main dish. Flaxseeds and chia were used in healthy drinks (Times of India, 2018).

FIGURE 7.4 Zoodles with garlic bread.

FIGURE 7.5 Grilled prawns with herbed rice and charred peppers, a nonvegetarian healthy food trend.

Going back to the basics, it became the mantra for the year 2018; be it cooking methods, farming, recipes, and eating habits, everything went back to the traditional way. Chefs went back to grandma style of cooking and the use of millets like jowar, bajra, and ragi was seen all over. Chef-created ragi cakes, bajra cookies, and finger millet biryani became one of the best and most popular dishes across Indian restaurants. The rise in the whole grain is a common trend and is seen growing over the last 2 years. Increased awareness about benefits of whole grains, global food crisis, and changing of the lifestyle and eating habits are the main reasons for them to gain popularity (Gandhi, 2018). Farmer's market became more popular in 2018, right from five-star hotels to basic restaurants, chefs preferred to buy vegetables from walking distances; this adds to the feeling of localism, almost every neighborhood saw a new farmer's market in its surrounding. This also helped the trend of back to basics, not only the hotels, but also the general public preferred buying vegetables and fruits from the farmer's market (Nayyar, 2018). The option for choosing food wisely has become a global phenomenon. People all over the globe are becoming environmentally conscious, and if their eating habits impact the environment, then they want to change the way they eat. The idea is to take little steps to save the planet; although imported lamb chops and salmon will continue to be on the five-star hotel menus, but more and more restaurants are offering local dishes. In fact, some of the restaurants are even offering Manali trout instead of imported fish. At Olive restaurant in Delhi, Chef Dhruv Oberoi has introduced indigenous leafy greens that were not even known to chefs, let aside general public. Jute leaves, Ceylon spinach, and green amaranth are some of the other examples (Ganguly, 2018).

Some of the specific trends that were seen in general include:

- **Knowing food:** the customers wanted to know about the purchases done by the chefs. they want to know from where their raw material is coming, and what impact does it have on the environment.
- **Edible flowers:** edible flowers were seen all over right from the starters to desserts. The use of flowers makes the dish look colorful and it healthy; however, this trend is so far restricted to five-star hotels only, but soon will catch up in another category of restaurants.
- **Detox diets:** increased lifestyle issues have not only helped dieticians to grow but have also lead to introduce detox food items in restaurant menus. items like green juice (Aloe Vera), chia pudding, tomato salad, sprout salad, and oatmeal are becoming common.
- **Fading of molecular gastronomy:** the concept of molecular gastronomy looked fading away from five-star hotels, but it started catching up in restaurants. the use of chemicals not only changes forms of the food, but also plays with the nutritive value of the dish. therefore, the concept looked fading away.
- **Fusion food:** with molecular food going away, the fusion food is back to the menu. the recipes were still prepared using local recipes and ingredients, but it is presented in a very modern way.
- **PrepPlated food:** most of the restaurants, even the best in the segments are going away with the platter to plate service. all preferred preplated food. this helps chefs to showcase their skills and at the same time helps to manage the restaurant better.
- **Online chefs:** Social media has been a boon for all the chefs. it is the easiest and fastest way to reach the audience. there were times when legends like Gulam Rasool, Imtiaz Quereshi, and Hemant Oberoi were limited to hotel kitchens. they could hardly reach masses, but with social media, even young chefs are able to make a mark.
- **Overall experience:** Patrons are looking for the overall experience. food remains an integral part, but the overall dining experience of a restaurant is what makes the customers coming back.

In general, 2018 ended with trends on fermented foods, vegetable-based pastas, kitchen gadgets, and regional Indian and Portuguese flavors dominated all year round. Mushroom, edible flowers, beetroot, zucchini, chia seeds, and millets became the most used ingredients in 2018. Home cooks and home style cooking kept growing, and the general trend of eating healthy foods (Fig. 7.5) and keeping themselves healthy was seen across all the food and beverage sectors. Farm to table concept, home pop ups, table of 4–6, and farm lunches were the growing trends (Ganguly, 2018).

FIGURE 7.6 Vegetable-based pasta sheets.

FIGURE 7.7 Beetroot and spinach ravioli.

7.6 THE WAY FORWARD

Anoothi Vishal (2018) predicts that "*Maa ke haath ka khana,*" i.e. home-cooked food (Fig. 7.6) from homes will shift to five-star hotels; be it the cooking style, portion size, and recipes, the mantra will be less is more, and dishes will be prepared by using minimum ingredients (Fig. 7.7). Chef Ritika Gill from Curry Singh, Gurgaon, adds that she serves simple home-cooked food in her restaurant; the recipes are from her mother and grandmother. The ingredients are procured from the local market and dishes are prepared with minimum spices, like the turnip is just cooked with only turmeric, cumin, and salt, similarly the other dishes. The butter chicken, dal makhni, and sarson ka saag that are served in hotels are not prepared the way it is done in Punjabi families; in hotels, they are overloaded with butter and cream, which is very unlikely at homes. Similarly in Mumbai, Chef Prateek Sadhu at Masque is preparing to get Kashmiri Pandits' food on the menu; the chef grew up in Kashmir eating home style Kashmiri food, and to get the authentic taste, he has tied up with local farms that grow vegetables as per his requirement. As a trend, he is also introducing a small tour for his customers in between the meals. He adds that this will help them feel like home; it will be exactly like he used to visit his mother's kitchen in between the meals. These types of trends will elevate the overall experience of customers. To take the home cook concept further, Chef Sadhu has hired a 68-year-old lady in his restaurant who specializes in making home style fish curry.

FIGURE 7.8 Live kitchen gaining popularity.

FIGURE 7.9 Traditional home style cooking.

In line with the trend, Chef Rahul Gomes of Pass Code Hospitality, in Goa, devoted his food to the extended family with whom he grew up. He opened up a restaurant in Goa, Pico's, where he is focused to use only Goan local ingredients and the entire kitchen will be run by only Goan female chefs. Similarly, there are other examples who seem to be experimenting with home style cooking (Fig. 7.8) and home style cooks, which are going to be the biggest trend of 2019. Chef Amit Chakraborti, executive chef at Banjara Mumbai, feels that live kitchen and live cooking (Fig. 7.9) will be the trend in coming years, but some of the trends which will be seen in the future include:

- Whole grains: They are there for 2 years in a row and seem to be dominating in the years to come; however, this year efforts will be to replace starches with healthy whole grains, millets, couscous, maize, and finger millet that will dominate the Indian food business.
- Magic ingredient: The year and years to come, it will be turmeric. it has already started gaining popularity as a magic ingredient. there are cafes in London offering turmeric-based drinks. similar concepts will be seen in India in the near future and years to come. the antibacterial and anti-inflammatory properties of turmeric will make it the magic ingredient.
- Beans and legumes: Beans and legumes will increasingly find their place in menus; these are rich in protein, and give energy so that

they become a good substitute for vegetarians. Apart from the local kidney beans, fava beans from Italy and other varieties from Africa and Peru might be seen in the menu.

- Substitute: Substituting the main ingredients with a healthy alternative is going to be the future trend, like almond and soy milk may replace cow milk, chickpea and red wheat flour may get replaced with the regular flour, pasta may be seen with chickpea or lentil flour, and goat milk may find its place in regular pasteurized milk.
- Crawling: This is a trend that is gaining popularity world over, but in India it still seems to be far away to be seen in the menus. although in some of the tribal communities, insects are eaten in various forms, but it is a rarity to be seen on elite restaurants in the near future.
- Flavored fats: Another interesting concept that may find a place in coming years is the use of flavored fats. the use of flavored oils was already in place, but flavored fats look fascinating, interesting to see the use of a variety of ghee and coconut butter chocolate.
- More powder: There was a trend of using powders in the last year and this is going to remain in place; in fact, it will grow even further. Some of the common flavors that continue to dominate include beetroot, coffee, turmeric, spinach, bell peppers with mushroom, and broccoli newly been introduced, and the focus remains on healthy food based powders.
- In-house: There was a time when almost all the preparations were done in-house. with time, the focus shifted and economics of food took over, but it looks like that some part of the in-house preparations are going to come back, especially the masalas, pastes, sauces, and dips that have a direct impact on taste and perception. for instance, there was a time when all hotels use to prepare fresh mayonnaise every single day, but today, there is hardly any hotel who is making on their own mayonnaise; everyone is outsourcing it. primarily, it is cost-effective and saves manpower. similarly, other items like salsa, hummus, cocktail, tartare, breads, croissant, etc. will see in-house preparations.
- Tangy foods: the acidity, salt, fat, and heat add the punch to the food, but so far Umami has dominated the taste buds, but with regional food gaining popularity, sour foods are bound to gain popularity.

- Ice creams: Typically it is vanilla, butterscotch, or chocolate ice cream that is most popular in Indian restaurants, but the trend is changing. fruit-flavored and dessert-inspired ice creams are already into the market; however, in future, these trends will shift to all natural and chemical-free ice creams. ice creams made from Tahini, feta cheese, cinnamon, peppers, etc. can be seen trending in future.
- Frozen desserts: The Italian gelatos are already in the market, but with the localization of ingredients, people are getting health and environment conscious. it seems the Indian gelato "Chuski Gola" crushed ice flavored with local fruit will gain popularity
- Natural sweeteners: Stevia is there for quite some time now, but has not able too quite dominate the industry. in future, especially in India, Honey and Jaggery are picking up. Big brands have already started pushing organic jaggery in market and honey was already there, but in coming years it will be the flavored honey from natural nectar that will dominate the sweeteners section of supermarkets and commercial kitchens.
- Kitchen garden: Space is one of the biggest constraints in bigger cities, but with the increase of rooftop and vertical farming, more hotels will start growing their own herbs and spices in future. the trend is already in, hotels like Taj, Oberoi, Aloft, and Holiday Inn Delhi have started nurturing fresh herbs and greens.

FIGURE 7.10 Sustainable kitchen garden.

Less is more will be an integral part of the cuisine trends in future. Chefs want to preserve the natural taste of the main ingredients. A chicken dish should taste like chicken not like cinnamon or cumin; however, the complimenting spices will be used but in moderation. This has resulted in more and more dishes in the menu that are seen with minimal ingredients. The focus of the customers is shifting toward environment-friendly practices. Even in the restaurant, customers are enquiring about the labels of the packed products, which were a rarity; reading of the labels by customers is a big indication of the inclination toward healthy and sustainable food (Fig. 7.10). This will also help customers understand their raw material and the source of procurement. Transparency in all the aspects will continue to be the trend in near future. The demand for plant-based diets will continue to grow; more people are looking for vegetarian and vegan diets; however, animal protein, especially white meat will still be an integral part of the menus. Gourmet food is going to be trending in the years to come. A report by Indian retail (2019) shows that the market for exotic food items, with unique flavors, fresh, and attractively packed stands at 155 million and is growing at a CAGR of 20%. The buyers are majorly in urban cities, have traveled widely, and understood the concept of gourmet meals; therefore, it is going to be catching up; however, the business will grow as the population and exposure of these cities grow (Indian Retail, 2019).

Learning for the traditional Indian food and eating habits, fermented foods are going to be the next big thing in food and beverage space. The increased awareness about the importance of keeping gut healthy and probiotic drinks helping the cause will lead to increase in adding fermented food and drinks in the menu. We don't see this as the trend of 2019, but the way its tone is being set, it is not very far. The Indian food industry does get some of the trends from the Western world and looking at the gaining popularity of cannabis in the United States of America and Canada. Chances are that it might be seen in menus in the future. The hemp seeds were legalized in the parts of USA and Canada that resulted in its growth; however, in India, it is still illegal to grow. However, there were incidences, like in Aahar, one of the biggest food festivals in India to showcase food trends. Canadian stalls were seen promoting and selling hemp seeds, but before this concept gets picked up in Indian food industry, it will have to go through policymakers. The Indian urbanization story is a little different from the rest of the world. People from small towns have migrated to the metropolitan cities, so the focus and development are seen only in the metropolitan cities. This has resulted in widening the Indian palate but at the same, they taste remain regional.

7.7 CHEF'S VIEW ON THE FUTURE OF FOOD

At the domestic circuit, Chef Sabyasachi Gorai believes that sustainability is not a trend it is going to stay, sustainability is here for us, we are here not for the sustainability, and it is rather going to save us (Fig. 7.11). it is the integral part of backward integration and will help farmers to sustain and in the process will save the environment. This will also help to support the cause of "*zero hunger*", which can largely be attained by reducing wastage and promoting the concept of food sharing. He further adds that the concept of "*Nose to Tail Eating*", started by Fergus Henderson in his restaurant St John in London, back in 1994, is going to be picking up in India, a British tradition that makes the best use of every part of the animal. This is exactly the opposite of some of the finest cooking of the world where the focus is on the use of the best cuts of the animals and rest is treated as carcasses used for making supplementary dishes. It promotes the use of the whole animal and a similar trend will be seen in vegetarian cooking, not only the main part but also the stem, flowers, and leaves will be used in cooking. This also means that chefs need to be creative with the offerings and need to research deeper for indigenous recipes and eating habits; apparently, some of the tribal areas in India do use all the parts of animals, not as a part of sustainability but because of limited access to other resources. Consumers are increasingly becoming health conscious. they think twice before making a choice, among Indian customers, eating healthy generally relates to vegetarian food, but nowadays even the meat lovers are talking about free-range animals or farm fed animals. the animals treated on antibiotics will be highly discouraged. With increasing awareness among customers about environmental practices, the customers may start questioning animal sourcing in the future (The Hindu, 2019).

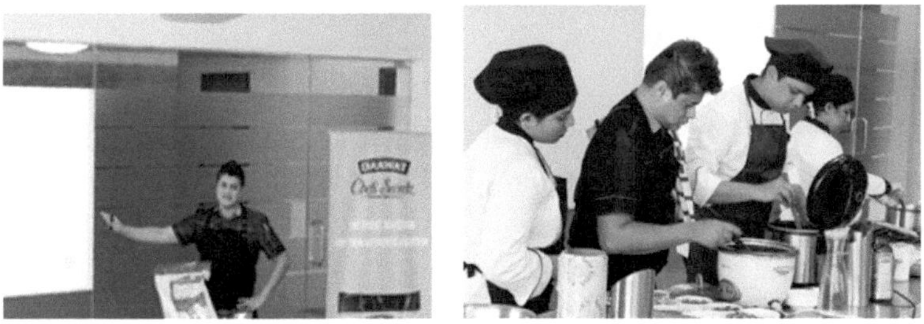

FIGURE 7.11 Chef Sabyasachi Gorai training and presenting lost Indian recipes.

Chef Garima Arora, an Indian chef in Bangkok, made the headlines last year after becoming the first Indian lady chef to win a Michelin Star. She currently runs a restaurant in Bangkok, GAA, and has previously worked with Chef Gordon Ramsay and Chef Gagan Anand. Chef Arora feels that chefs will look inward for inspiration in the future, rather than looking and following market trends. chefs in the future will look into their origin, their culture, and their cuisine. She feels that sustainability will remain the core, but commercial kitchens are inherently sustainable; chefs are always on toes to monitor the wastage and keep the food cost under control. therefore, it a part of their regular job, but restaurant kitchens, restaurants, and hotels are a very small part of the bigger picture. they can only contribute within in the limits, so the scope of these practices needs to be discussed and taken further (The Hindu, 2019).

Chef Vicky Ratnani feels that there will be a rise in the use of ancient grains. People have become extremely health conscious. They have realized the importance of eating whole foods and organically produced without any chemicals. Even the use of genetically modified seeds is discouraged. Grains like a variety of millets, amaranth, sorghum, and nachni will be used generously; these grains have become a preferred choice because of accessibility, health benefits, and cost effectiveness. A few years back, blogging was a mere hobby and neither the customers nor the restaurateurs took them seriously, but with regular quality and consistency, it has become a big trend today. Karan Tripathi is one of such bloggers; he feels that health and healthy eating is going to be the focus. Therefore, the use of microgreens will be the trend in future. It not only makes the dish look good but also adds nutritive value to the dish. Vegan will continue to grow and vegetarian options will have more offerings rather than traditional salads.

7.8 CONCLUSION

One thing is very apparent that the inclination of customers, in general, is toward healthy eating. People are looking for the sustainable food habits and how their buying behavior is going to impact the environment and society at large will help decide their buying pattern. Climate change is not only impacting the environment but also impacting the food business. We need to constantly talk about the issues and challenges. The discussions on bigger issues need to come out of formal setups of seminars and conferences, and at every level it needs to be addressed. If it is impacting food business, then chefs, bloggers, food critiques, and farmers need to write and talk so that the effect is visible at the grass root level. Sustainability is the need of the hour, not just domestically but even internationally the concept of slow food, good, clean, and fair food is picking up. Increasingly, people are growing and consuming local ingredients and are concerned about the impact of their eating habits on the environment. Organizations like tasting India symposium aim to preserve the local taste and encourage the use of heirloom ingredients.

The increasing interest of patrons for transparency will continue to put pressure and chefs will create dishes with the best ethical practices. Another differencing factor will be the way these practices are communicated to the customers; it is to build the perception among the customers and follow what is being promised. However, if you look at the other side of the coin, the biggest challenge for the chefs is to attract good people and retain them. The demand for employees is also increasing by the time; providing a good environment, better wages and other perks, and at the same time the outlet needs to be profitable is going to be the biggest challenge. The entrepreneurs who are able to set the right balance between customers, employees, and profitability will be able to sustain for a longer period of time. Innovation in menus and training of the staff will be helpful to some extent. The staff should feel motivated to learn and deliver to the set standards.

KEYWORDS

- **sustainability**
- **healthy food**
- **future trends**
- **future of food**
- **chefs predictions**
- **trending foods**
- **future cuisine**

REFERENCES

Achaya, K. T. *A Historical Dictionary of Indian Food*; Oxford University Press: New Delhi, India, 1998.

Appadurai, A. How to Make a National Cuisine: Cookbooks in Contemporary India. *Comp. Stud. Soc. Hist.* **1988,** *30* (1), 3–24.

Anoothi, V. 2019's Biggest Food trends may see chefs bringing their own culinary cultures to restaurant kitchens. [Online] 2018. https://economictimes.indiatimes.com/magazines/panache/2019s-biggest-food-trend-may-see-chefs-bringing-their-own-culinary-cultures-to-restaurant-kitchens/articleshow/67302687.cms?from=mdr (accessed Jan 10, 2019).

Gandhi, N. Restaurants and food trends. [Online] 2018. https://magicpin.in/blog/restaurant-trends-2018/ (accessed jan 4, 2019).

Ganguly, T. Food Trends 2018: The ones that will matter the most for Indians. [Online] 2018. https://food.ndtv.com/food-drinks/food-trends-2018-the-ones-that-will-really-matter-to-indians-1795416 (accessed Jan 8, 2019).

Goswami, S. Top seven trends that will hit the Indian market. [Online] 2018. https://www.indiatoday.in/lifestyle/food/story/food-trends-2018-india-shell-less-tacos-plant-based-mushrooms-regional-cuisines-no-waste-lifest-1119513-2018-01-01 (accessed Jan 2, 2019).

Indian Retail. 6 consumption trends that will change the Indian food Industry. [Online] 2019. https://www.indiaretailing.com/2019/01/24/food/food-grocery/6-consumption-trends-that-will-change-the-future-of-indian-food-industry/4/ (accessed Feb 3, 2019).

Jaffrey, M. *A Taste of India*; Pavilion: London, 1986.

Nandy, A. *The changing popular culture of Indian Food: preliminary notes*; Sage Publication: Thousand Oaks, London, Vol. *24*, No. 1, 2004.

Mehta, S. Top Chefs on India predicts food trends in 2017. [Online] 2018. https://www.mansworldindia.com/fooddrink/top-chefs-india-predict-food-trends-expect-2017/ (accessed Jan 1, 2019).

Nayyar, A. These food trends will rule 2018. [Online] 2018. https://www.restaurantindia.in/article/These-Food-Trends-will-Rule-2018.10482 (accessed Jan 7, 2019).

The Hindu. [Online] 2019. https://www.thehindu.com/life-and-style/food/sustain-retain-maintain-the-food-chain/article25899136.ece (accessed Jan 2, 2019).

The Restaurant Times. [Online] 2018. https://www.posist.com/restaurant-times/features/7-restaurant-trends-watch-2017.html (accessed Jan 1, 2019).

Times of India. [Online] 2018. https://recipes.timesofindia.com/articles/features/top-food-trends-youll-see-in-2018/photostory/62350342.cms (accessed jan 6, 2019).

CHAPTER 8

Integrating Backward to Achieve Culinary Sustainability

ABSTRACT

The chapter will discuss the need of backward integration to achieve culinary sustainability; it will explore the impact of sourcing local ingredients on the entire food chain. The role of, and, impact on farmers, retailers, chefs, and finally on consumers will be discussed. The concept of farm to fork will remain the core of discussion, with emphasis on organic farming, sustainability, and the need of backward integration. Some of the hotels, restaurants, and chefs from India and their views on integrating backward are discussed; further discussion is on their role in promoting the need of sustainable practices. Chef's knowledge on farming and its impact on consumers and business, play a crucial role in spreading awareness. As restaurants are increasingly using local organic ingredients, the term "farm to fork" is commonly seen in menus and other promotional campaigns; however, higher cost and credibility of the organic ingredients available remains the biggest challenge in gaining popularity and in making the backward integration sustainable.

8.1 INTRODUCTION

The theory of backward integration is closely linked with the concept of "farm to fork." While studying backward integration, it becomes imperative to focus on the distance that food travels from farm to table and all the aspects in and around the traveling of food are covered in the concept of "farm to fork." The concept of "farm to fork" is gaining popularity world over; there is no specific definition to it, but, it covers the journey of food from farm to table. Different authors have different definitions and assumptions; some believe that it means the food will travel directly to the restaurants or end users from farms without any middlemen, some believe it to be organically produced or food, that is seasonal and regional, but in essence, it is the food that is grown and consumed within the region, produced with environment friendly and sustainable practices and will help contribute in the local economy (Brain et al., 2015). For a chef to serve food on the concept of farm to fork, it is not mandatory to abide by all these rules; they can follow all or a combination of one or more rules, but, primarily, the differentiating factor between a normal and a farm-to-fork restaurant is the way food is sourced and its impact on society in general and the environment in particular.

While investigating the possibility of achieving culinary sustainability through backward integration (Fig. 8.1), food sourcing remains the most important aspect. In a restaurant; be it stand-alone or of a star category hotel, the purchase decisions are with the chefs; in bigger establishments, sous chefs take decisions in consultation with the purchase manager and in smaller setup, executive or the head chef is responsible for all purchase decisions. Typically the buying decisions for raw materials are based on taste, quality, freshness, and regular availability and the impact of the decision of the environment and other social factors are generally not considered (Curtis and Cowee, 2009). For more than over a decade, the demand for eating out is increasing; people are preferring restaurant food over home-cooked food, with the gaining popularity of the local food movement, the farm to fork will also become popular (Scheule and Sneed, 2001). However, in the Indian context, the majority of the people are still eating home-cooked food, but, in bigger cities, with the increase in disposable income, expanding nuclear families, and women opting for outdoor work, the restaurant business is flourishing. The integration of restaurants with local farmers has a great potential to support the local economy; promoting local ingredients will reduce middlemen, cut down the traveling time, increase jobs, and be able to provide healthy seasonal food options; however, procurement from local farmers can be little expensive; they do not grow in larger quantities and therefore, cannot leverage on the economy of scales and also are not equipped with the latest gadgets, so the time and effort increases, which results in lower yields and higher prices. This is one of the main reasons for restaurants concerned about environment and sustainable practices pricing their menus higher than normal restaurants.

The sustained use of locally grown ingredients is very important in achieving culinary sustainability. Local food systems might not be a sustainable model within themselves, but, they play a crucial role in the sustainability of the entire food chain. While we are discussing the term culinary sustainability, the term sustainability itself has multiple meanings; most of the people relate sustainability with environmental sustainable practices; water conservation, responsible use of electricity, promotion of solar energy, air and water pollution, and recycling and reusing of waste are topics of general discussion, these environment sustainability topics are essentials but focusing on culinary sustainability will not only cover environmental best practices, but will also cover the economy and society as a whole (Trivette, 2012). With the industrial revolution the food business changed more than ever, production and demand for prepacked, processed, and convenience food increased. The developments in the agriculture sector changed the consumption pattern; year-round availability of ingredients took seasonal menu planning away, the concept of having menu items on season started fading. On a bigger scale such developments have negative impacts on aspects like food safety, food security, increased life order diseases, animal safety, ecological disorder and most of these aspects are directly linked with the modern agricultural practices and policies (Counihan, 2007).

8.2 NEED FOR BACKWARD INTEGRATION

Backward integration will have several environmental benefits in response to the industrialization of agricultural practices. Sourcing of local ingredients will reduce; transportation cost, time, carbon emission, and most importantly will supply fresh ingredients to help build the local economy stronger. Sourcing locally is closely linked with the supply of organic produce; the practices for organic produce treat soil and environment ethically, the idea is to minimize and gradually eliminate the use of harmful chemicals in soil. This will not only help in the near future by providing healthy food options to the society but will also retain and increase soil fertility to be used for longer period of time.

FIGURE 8.1 Eminent chefs from India discussing on the need for culinary sustainability.

8.3 FOOD TRAVEL

The rise of the international cuisines in India resulted in increased traveling time, more and more food traveled from the country of origin. For instance, Gung the Palace, a Korean restaurant in Delhi NCR; even today imports most of its supplies from Korea, as a result, the menu prices are higher (Fig. 8.2). Not only the traveling time and distance; to keep them fresh and available for a longer period of time they need to be stored properly, but the modern-day storage techniques also use fossil fuels which negatively impacts the environment, releases greenhouse gases, increases temperature, and is responsible for the depletion of the ozone layer. Issues and challenges like these gave rise to the concept of food miles, the distance that food travels from its place of production (farm) to middlemen to the place where it gets cooked and finally served (restaurant) (Iles, 2005); this also implies that consuming locally produces food is more environment friendly. The greenhouse emission not only takes place through the transportation of the food items, but the production of food also contributes to the emissions. A study jointly conducted in India by National Dairy Research Institute and an Indian Institute of Wheat and Barley Research compiled the statistics and compared the emission of grains like wheat, barley, fruits, vegetables, and the livestock products like milk, eggs, chicken, and other meats. It was found that livestock and rice were the main sources of greenhouse emissions in India. The country's average for rice was 5.65 kg CO_2eq kg^{-1}, 45.54 kg CO_2eq kg^{-1} mutton, and 2.4 kg CO_2eq kg^{-1} milk. They also concluded that the production of cereals (except rice), fruits, and vegetables emit lower amount of greenhouse gases as compared to meats and meat-related products. It also implies that a general inclination in the country towards nonvegetarian food may seriously harm the environment and vice versa. The adoption of locally produced items will greatly help in achieving culinary sustainability in particular and environment sustainability in general. The onus to reduce food miles and adopting sustainable practice is on all the stakeholders; it includes farmers, retailers, chefs, and the culinary profession.

FIGURE 8.2 Gung The Palace is known for serving Korean food in India.

8.4 ORGANIC FARMING

Organic farming is associated with healthy farming practices; it discourages the use of chemicals and fertilizers in farming. But looking at the gap between demand and supply, farmers are increasingly using modern techniques. However, traditionally, organic farming was limited to consuming food close to its production (Belasco, 2007). According to the world of organic agriculture report 2018, more than 30% of the world's total organic producers are in India, yet it only accounts for 2.59% of the total organic cultivation. Low yield and increased cost are two of the main factors that discourage farmers from adopting organic practices. Sikkim was declared as the first organic state, but during the transition, farmers faced issues and challenges with low productivity and increased plant diseases. They requested the government to provide sufficient resources and teach them from dealing with the adversities. Similarly, in Rajasthan, 98% of the farmers are well aware about the ill effects of prevalent chemical and fertilizer based farming, but low yield and nonavailability of the organic resources discourage them from going organic. According to ASSOCHAM 2018 report, special farming training, inventory cost, packaging, promotion, logistics, and small distribution network results in increased pricing. Even if, after all the efforts the produce reaches the market, then also because of the higher prices customers are not willing to buy organic. According to the report, monthly expenditure of a common household rises from Rs. 1200 to Rs. 1500. Moreover, the current certification in India is cumbersome and is expensive too. After a long protest by farmers in India, FSSAI (Food Safety and Standards Authority of India) is considering a change in the norms which may help increase the interest of the farmers to go organic (Pandey and Sengupta, 2018). While the government has taken preventive steps by introducing *Parampragat Krishi Vikas Yojna* and have made the certification free; farmers and state government are not been able to utilize the scheme efficiently, except for the states of Karnataka, Tripura, and Orissa, more than 50% of the funds remained unutilized. In essence, organic farming does have the potential to introduce environmental sustainable practices and help backward integration become an integral part of culinary sustainability; it needs to be understood from a bigger picture of the ecosystem as a whole (Fig. 8.3).

FIGURE 8.3 Fresh farm produce getting ready to be supplied.

8.5 ECONOMIC VIABILITY

It is always good to have systems that are environment friendly, but at the same time they need to have economic viability to be able to sustain. Looking at the challenges that we discussed so far, it is becoming increasingly difficult for the small farms and farmers to be in business; increased pressure from bigger retailers are pushing them out of the business. The introduction and significant increase in supermarts have put more pressure on the supply chain; they have better access to the producers and are able to source from one to two big suppliers, they get the advantage of economy of scale and make it even more difficult for the local farmers to show their presence in the market (Halweil, 2002). This concept is not only true for the supermarts, but even for the bigger hotels, restaurants, canteens, cafeterias, hospitals, and institutions it holds equally true. Small farmers find it difficult to reach such institutions; therefore, gets discouraged to opt for environmental healthy practices. One of the solutions to such problems is to get cost-effective agriculture practices and tie up with agricultural organizations that will help them provide cost-efficient solutions. One such organization is Indian Agricultural Research Institute, Pusa, New Delhi; they provide education on the field of crop improvement, plant protection, basic sciences, and natural resource management through free seminars and collaterals. So, the government is taking initiatives to promote local and seasonal ingredients and make organic farming an economically viable option, but at the same time, if the hotel industry is not been able to encourage the use of local items and consumers are not made aware about farm to table, it will become increasingly difficult for the local farmers to sustain. On the other hand, slow food in India is gaining popularity; it preaches good, clean, and fair food. Good for the consumers, clean for the environment, and fair for the growers; hotels and restaurants have already started offering menus based on the ethos of slow food, this will greatly help in achieving culinary and economic sustainability, as purchasing local ingredients will help circulate the capital with the local region, instead of traveling to the multinational national companies headquarter elsewhere (Halweil, 2002).

8.6 SENSE OF RESPONSIBILITY

One of the basic needs to look for backward integration is to realize the sense of responsibility; the responsibility that we have for the people and for the society, it is the prime responsibility of all the hoteliers, chefs, and restaurant managers to take care of the customers. The hospitality needs to be deep rooted; welcoming guest with a smile, taking care of the basic needs, delicious food, courtesy, empathy, etc. are very important, but sense of responsibility goes beyond the basic expectations; it addresses the promise to customers to feed with a variety of safe, healthy, and nutritious food; similar to what slow food has to offer; good, clean, and fair food. The promotion of local food system also promotes the greater social connection between the consumers and producers, which in turn helps the business in multiple ways (Ikerd, 2007). At the same time, some people do not promote the idea of a local food system, simply because it lacks convenience; they want food in their doorsteps, as and when they want, for them seasonality also reduces their options to explore options and it becomes monotonous for them to have repeatedly the same ingredients again and all over again (Schneider and Francis, 2005).

8.7 PRACTICING BACKWARD INTEGRATION IN INDIAN

The concept of farm to fork is not new; it started in California in the 1960s and gradually got the momentum to reach the rest of the world. Much of the credit goes to Chef Alice Water, who is believed to get the popularity of the concept to where it is today. After moving to California from France, she opened her restaurant in 1971 and realized that the freshly procured vegetables are much more flavorful than the vegetables received through the regular supply chain. She got acclaimed globally for the food that she was serving at Chez Panisse. So, along with the restaurant, not only Chef Water became popular, even the concept of using fresh ingredients from the local community also gained popularity. By 20th century it got popular world over and in Asia, it became much popular during the 21st century. However, as we discussed in our previous chapters, traditional Indian food was by default based on the concept of using freshly available food. Even today, unlike most of the Western countries; Indian households cook meals three times a day; the reasons could be many, but in essence, the tradition of using and cooking fresh ingredients is still prevalent in most parts of the country. However, popularizing the concept and looking for the need of backward integration only started recently, it got popular as a concept when chefs from restaurants and hotels started looking at it and started getting it featured on the menu and promoting it at various platforms. Today, most of the upscale restaurants have a full menu based on environment-friendly practices or served them in parts. In fact, they project and promote it as their USP and with customers getting sensitized, it seems to be working. While the concept of fresh ingredients and ethos of slow food prevailed in Europe by the efforts of the likes of Noma; it was evident to become popular in other parts of the world, while some chefs and restaurateurs are genuinely working on it, the rest are just using it as a marketing gimmick and riding on the wave (Vishal, 2016).

FIGURE 8.4 Chef Achintya Anand in conversation on the importance of backward integration.

The journey of farm to fork from the restaurants of California to the suburbs of Delhi has been very inspiring. Chef Achintya Anand, a culinary graduate from Le Cordon Blue, Australia, got his inspiration to work with fresh ingredients while working in Aquavit, New York. After coming back to Delhi, from a Michelin star restaurant in New York, Chef Anand (Fig. 8.4) joined “Tres,” a modern European restaurant and started exploring with fresh ingredients and microgreens; it is then he started visiting restaurants and hotels to check the demand and supply for microgreens and in the process he became the most popular supplier of the microgreens in Delhi NCR. The demand rose to a level, where Chef Anand converted into a full-time farmer growing and supplying exotic greens, herbs, and other vegetables. Some of his prime customers include The Oberoi’s, Leela Palace, ITC Hotels, Olive, etc. (Brown Paper Bag, 2019). Chef Sujan Sarkar is another Indian chef who operates a restaurant in New York and believes in using freshest of the ingredients with traditional recipes, but the dishes are presented with a modern touch. When you see food items like “Ek Bar Granola Bar,” you know you are in “Ek Bar”; a modern Indian cuisine restaurant in New Delhi by Chef Sarkar. The ek bar granola bar is a modified version of Bengali jhal moori with granola nuts and dhai bhalla sorbet. Similarly, sausage thali is an assortment of Indian style sausages served with pickle and mixed chutneys, not only food, even drinks are served with a twist. "Murabba Mule" is a fusion of Moscow mule and Indian murabba.

FIGURE 8.5 Chaupal; a platform to get chefs, farmers, hoteliers and customers together.

The backward integration could not be possible without the support of the farmers; not only the chefs, even farmers those who are not from culinary background are also taking keen interest in growing exotic vegetable and fruits (Fig. 8.5). One such example is Saurabh Khanijo, who owns a restaurant in Delhi, but come from a family of poultry farmers. They had to close their poultry farm because of low demand and high maintenance cost. However, with the popularity of backward integration in restaurants to help achieve culinary sustainability, the demand for fresh produce increased significantly; therefore, they converted their land to vegetable farms, now they grow vegetable without the use of any chemicals or pesticides. Even basic vegetables like potatoes, tomatoes, and onion are grown without any additives. To maintain the ecosystem, they have around 200 chickens which produce free-range eggs; the color, taste, and size of the eggs are very different from regular eggs; egg yolk is absolutely orange because of the natural feeds from the farms. Chef Jamie Oliver publically announced in one of his culinary shows that the herbs used in the cooking show are all from his own farms; in fact to get the best results of his recipes, he even motivated and encouraged spectators to use fresh ingredients. Chef Lokesh Swami is another culinary professional who is practicing backward integration; he works very closely with Chemon farm in Delhi to get fresh produce, for him getting vegetables from Chemon farms is like traveling of vegetables for only 10 km. For a chef, it is important to use green leafy vegetables while they are still young and getting them from a nearby farm ensures the highest quality, so lettuce, kale, spinach, rocket, amaranth, etc. that are received from Chemon farm are far too superior in quality and freshness than any of the regular vegetable supplier. Chef Swami works so closely with the farmers that in the month of August he sits with farmers to check the availability in October and they plan their menu accordingly.

FIGURE 8.6 Chefs on farms, an example of Zero Miles.

The Bombay Canteen in Mumbai is known to serve healthy Indian cuisine with a modern twist, while compiling the menu at the Bombay Canteen, Chef Zacharias visited 8–10 Indian states; he traveled to Punjab, West Bengal, Sikkim, Tamil Nadu, Meghalaya, and Gujarat to look for ingredients, recipes, and stories behind the traditional food served in those states. After returning back to Mumbai, he followed the same recipes. However, the ingredients are sourced from local farms and farmers' market. He believes in getting maximum yield from any of the vegetable, for instance, in pumpkin; the fruit is cooked and an accompaniment to the main dish, so are the leaves and stem, he uses flower of pumpkin to make fritters. At The Bombay Canteen, they have introduced an Indian weekly program, where every week they select an Indian grain and work separately to research on its origin, ways of using and current usage and then towards the end of the week they all share their research and decide if it can fit in the restaurant menu. Chef Atrie in Calcutta runs a local gourmet; she is one of the chefs who are focusing more on the nonvegetarian options available locally. Fenwick bazaar in Calcutta is a true Eurasian market known for serving good quality of meats, like Panchetta and she is a regular visitor of this market.

The backward integration has helped get good quality food, saves cost, and helps is getting repeat business. Moreover, it gives you a sense of pride by serving fresh and healthy ingredients straight from your farms (Fig. 8.6). Not only chefs, even other professionals are increasingly promoting the need of backward integration. One such example is Vineet Bansal, an engineer from Gurugram turned his hobby into terrace farming. After the first harvest he realized the difference in the taste of vegetables grown at home and vegetables bought from the market; since then, he grows his own vegetables to the extent possible. Even celebrities are also helping in promoting the concept of backward integration; Australian cricketer Mathew Hayden has converted the majority of his land into a farm and uses the produce to experiment with the recipes.

FIGURE 8.7 Sandeep Munjal addressing the need for backward integration.

8.8 BENEFITS OF BACKWARD INTEGRATION TO HELP ACHIEVE CULINARY SUSTAINABILITY

There are multiple benefits of making backward integration stronger, whether it is an operational fully-fledged restaurant or relatively new setup; it is worth understanding the value proposition of locally sourced food. At the surface level we think of fresh, tasty, and nutritive food, but in business terms, it has much more to it. In principle integrating backward will ensure fresh produce, locally grown, supplied straight from the farm to kitchen. Let us look at some of the aspects that will result positively while integrating backward (Fig. 8.7).

8.9 CHEMICAL FREE INGREDIENTS

In the modern time where demand seems always more than the supply, farmers and producers are under continuous pressure to increase yield; therefore, to meet the need, they all are tempted to spray chemicals and add growth regulators in the hope of better yield. In a short time, it is difficult to understand the ill effects of the harmful chemicals being used in various vegetables and fruits, but in the longer run, symptoms are visible. Integrating backward will ensure fresh ingredients right from the farms. These are naturally grown, free from chemicals, and gives peace of mind to chefs and customers.

8.10 ECONOMIC VIABILITY

While looking at the backward integration, chefs and managers are constantly looking at the benefits that it may have on the bottom line. However, it will also help local farmers financially, and this in effect will give motivation to the farmers to grow more varieties of exotic vegetables, with environmentally best practices. Moreover, the money will remain within the community, making the local community economically stronger.

8.11 ENVIRONMENT PROTECTION

Food that is grown locally will not have to travel long distances; therefore, either will not add on to the global warming or the impact will be very minimal. On the other hand, if fruits and vegetables food are to be transported, then chemicals are added to fruits and vegetables along with the packaging which results in carbon emission, also the vegetables are plugged raw and the packed with chemicals to ensure freshness for customers. Such kind of raw material will always be of inferior quality as compared to freshly procured.

8.12 EDUCATING YOUTHS

Chefs and managers have realized the importance of backward integration, but it will be more helpful if they are able to teach it to the youngsters and get the same implements in college and school level. The trend and need for backward integration is prevalent in restaurants, but the concept is not seen in school and university canteens, looking at the health benefits and increasing awareness about eating healthy, one should start teaching kids and get it implemented at the root level.

8.13 INQUISITIVE CUSTOMERS

Customers are getting more conscious about their eating habits, they know, what to eat and when to eat; however, customers increasingly want to know about the source of procurement of raw material used in making of their dishes. It is not only the calories or the amount of fat present in the food that customers are willing to know; they even want to know, if the entire procurement process was environmentally sustainable. For instance, while knowing about the meat, customers want to know if it was free meat without any steroids or chemicals. The procurement process and source of procurement are two of the main components of integrating backward. So in such a scenario, restaurants offering meals based on best backward integration will have an added advantage and be able to survive in the long run.

8.14 CONCLUSION

India is the second biggest agricultural producer in the world after China so, it deserves to get attention on the farm to fork and backward integration. The concept is already gaining popularity in the metropolitan cities of India. Chefs are increasingly using fresh and locally produced ingredients, and customers not only want to know about the dish and the recipe but also want to know about the procurement process of the raw material used. This looks like a two-way process where both; the restaurateurs and the customers are moving in the same direction, chefs and managers are looking for traditional, lost ingredients, and recipes, which can be sourced regionally and seasonally and customers are demanding fresh and healthy produce. Popularity of concepts farm to fork and the need for backward integration adds a new dimension in menu planning, typically; demand, trends, future, manpower, skills, etc. are considered while planning a menu, but, focusing on backward integration will add another feature to be considered while planning a menu. The chefs need to meet farmers more frequently, they need to sit together and discuss what is growing and how it is grown, for restaurateurs and chefs who are looking for culinary sustainability through backward integration; farmer's sowing plan will become an integral part to consider while planning a menu. On the other hand, feedback and experiments from chefs will identify products and raw materials to be worked on, and depending on the feedback from chefs to farmers, the soil treatment will be kept in mind when sowing the same vegetables the next time. So, basically, the soil planning for farmers and menu planning for chefs will go hand in hand, this will gradually improve the quality of soil and will result in achieving overall sustainability of the community.

There are multiple reasons for chefs who are not ordering food items locally. However, consistency, quality, regular follow-ups, improper supply chain, lack of awareness about the supplier, and their credibility remains the biggest challenges. Some of the restaurants are aggressively following backward integration, but to make it a mass movement, formation of a community is required. The community will comprise of leaders in the business, local farmers, restaurant owners, chefs, local government bodies and discuss strategies for the way forward, else it will become difficult for small scale and budget restaurants to adopt and gain from backward integration. The focus of the community needs to be on aspects like:

- Understanding the need of the restaurants, especially in the urban market and guide farmers to grow crops as per the demand
- Chefs from hotels and restaurant, those who have already implemented the concept and are gaining popularity, need to train and guide other restaurants and chefs
- More farmers market within the reach of every society to be opened and create awareness about all the aspects of backward integration; so that the concept spreads in households rather than limiting it to only hotels and restaurants
- In rural areas, the concept of fine dining is not very popular; therefore, the focus has to be on the need of backward integration of

households and local farmers; establishing a local farmers' market will help attain backward integration to a large extend

- Discussions can be initiated on integration between home delivery applications and local farmers, so that, the delivery company has access to farmers and need for buying form a retailer or middlemen could be eliminated
- There is a strong need for hosting meals and food festivals within the farms, this will help create awareness and will also act as a platform for businessmen and farmers to come to a common ground
- Building of subjects and modules on sustainability, backward integration, and best practices, especially in culinary and hospitality schools will help create awareness, teaching at the base level will be an ideal way to communicate with the entire fraternity

In essence; backward integration endeavors in bridging the gap between the farmers/producers and end users, in case of commercial setups, chefs are the end users of the raw material; therefore, it is important for chefs to have good knowledge on farming, farming practices, seasonality, and environmental best practices. Chefs, overall knowledge on the importance of backward integration will have multiple impacts on culinary sustainability in particular and a healthy and safe environment in general. It will help reduce the food miles, resulting in improved taste, texture, and aroma of the raw material used. It will also allow embracing the regional and seasonal food, by accepting local ingredients the local community feels connected and such activities are useful in the overall survival of the community ecosystem. It will help nurture small business, will provide financial support to the local community, and promote alternate food system. To complete the backward integration and achieve culinary sustainability, chefs, restaurant managers, and policymakers need to implement and monitor best practices at all levels of from farm to fork, they even need to go beyond the plate and any leftover food needs to be segregated and should be decomposed responsibly. This will help complete the food cycle and will go a long way in achieving overall sustainability. Chef's knowledge of farming and its impact on consumers and business play a crucial role in spreading awareness. As restaurants are increasingly using the local organic ingredients, the term "farm to fork" is commonly seen in menus and other promotional campaigns. However, higher cost and credibility of the organic ingredients available remains the biggest challenges in gaining popularity and in making the backward integration sustainable.

KEYWORDS

- **culinary stainability**
- **farm to fork**
- **backward integration**
- **zero miles menu**
- **food travel**
- **organic farming**

REFERENCES

Belasco, W. *Appetite for Change: How the Counterculture Took on the Food Industry*. Cornell Paperbacks: Ithaca, New York, 2007.

Brain, R.; Curtis, K.; Hall, K. Utah Farm-Chef-Fork: Building Sustainable Local Food Connections. *J. Food Distrib. Res.* **2015,** 46 (1), 1–10.

Brown Paper Bag. [Online] 2019. http://brownpaperbag.in/delhi/food-and-drink/meet-and-green-with-achintya-anand-of-krishi-cress/ (accessed Feb 14, 2019).

Counihan, C. *Food Politics: How the Food Industry Influences Nutrition and Health. Food in the USA: A Reader*. Routledge: New York, 2007. Revised ed. University of California: Berkley.

Curtis, K. R.; Cowee, M. W. Direct Marketing Local Food to Chefs: Chef Preferences and Perceived Obstacles. *J. Food Distrib. Res.* **2009,** 40 (1).

Halweil, B. *Home Grown: The Case for Local Food in a Global Market.* Worldwatch Institute: Washington, DC, 2002.

Ikerd, J. "On Defining Sustainable Agriculture." North Carolina Sustainable Agriculture Research and Education Program. 2007, [Online] 2019. http://www.sustainable-ag.ncsu.edu/onsustaibableag.htm (accessed Feb 10, 2019).

Pandey, K.; Sengupta, R. India has Highest Numbers of Organic Producers but all of them are Struggling. [Online] 2018. https://www.downtoearth.org.in/news/agriculture/india-has-the-highest-number-of-organic-farmers-globally-but-most-of-them-are-struggling-61289 (accessed Feb 2, 2019).

Scheule, B.; Sneed, J. From Farm to Fork: Critical Control Points for Food Safety. *J. Nutr. Recipe Menu Dev.* **2001,** 3 (2).

Trivette, S. A. Farm-to-Fork: Understanding Locally-Oriented Farm-to-Vendor Food Systems: Access, Boundaries, and Power-Relations. 2012. https://scholarworks.umass.edu/open_access_dissertations/670

Vishal, A. Restaurant Ride the Farm to Fork Wave. 2016, [Online] 2019. https://economictimes.indiatimes.com/magazines/panache/restaurants-ride-the-farm-to-fork-wave/articleshow/51960397.cms (accessed Feb 10, 2019).

World Organic Agricultural Report. [Online] 2018. https://shop.fibl.org/CHen/mwdownloads/download/link/id/1093/?ref=1 (accessed Feb 2, 2019).

CHAPTER 9

Street Food and Commercial Expression in Food and Beverage Space

ABSTRACT

This chapter examines the scope of street food in India with issues and challenges faced by the vendors. Cases from the metropolitan cities of India will be discussed further. The food and beverage space in Indian metropolitan cities is growing more than ever. There was a time when fine dining and speciality restaurants were restricted to five star hotels but with urbanization and expanding corporate culture, stand-alone restaurant sector has seen tremendous growth. In Indian cities, urbanization, increasing disposable incomes, and nuclear families have led to increase in the demand for food and beverage business. Simultaneously, it has also impacted the growth and demand for street food; therefore, with every passing day, street food is becoming an integral part of the society. The initial setup for vending carts doesn't require a lot of capital and regulatory bodies monitoring these stalls are also negligible; therefore, more and more people are entering into the bandwagon of street food in India. The concept of chef manager with restaurants specializing in a particular cuisine is mushrooming in tier I and II cities of India. With the growing organized food and beverage industry, the demand for street food is also increasing. India is always known for serving street food, although, the offerings have transformed over the period of time; however, some traditional food items are still prevalent amongst customers. The cultural diversity of India is also visible in the street food that it has to offer. Through the discussion in the chapter, we will look at various offerings from across India with focus on the metropolitan cities.

9.1 INTRODUCTION

India has several options to offer when it comes to street food; it is difficult to define Indian street food in a few words. With every state in India, culture, tradition, language, preferences etc. change. Similarly, street food also changes with every mile. In contrast to the mid-segment restaurants, the street food is economical, freshly prepared but comes with other challenges. Typically, any food which is ready to be consumed and is offered by the vendors alongside streets is defined as street food (Muleta and Ashenafi, 2001). In India eating street food is much more than only eating; it is an experience that needs to be felt. It offers local specialities with regional influences which are generally not available in other countries. Currently, the turnover for the street food in India is estimated to be 30 million rupees every day. According the Food Safety and Standards Authority of India (FSSAI), 2019 Report, there are around 10 million vendors in Indian streets, out of which 2 million are food-related vendors (Fig. 9.1). The numbers look fascinating but the entire street food sector is highly unorganized. There are efforts being taken by the government bodies but it remains unorganized. Although there are basic food norms being set up by FSSAI but there is no common body that operates or regulates these vendors; therefore, there is high level of uncertainty. The street food in India is considered to be spicy, oily, and unhealthy and is served in an unhygienic condition, which is one of the prime reasons for it not getting popularity amongst international tourists (Lonely Planet, 2016).

There are multiple reasons for the street food vendors for not being able to serve the food under hygienic conditions. In general, food vendors are from poor background, they lack education, have no knowledge on Hazard Analysis and Critical Control Points (HACCP) or hygiene standards, no sense of keeping the environment clean, limited access to food preparation and service tools, and lack basic awareness on basic hygiene standards (Janie and Marie, 2013). There are numerous researches conducted on street food; however, one of the things which is common is poor hygienic conditions for the service of food. Inadequate personal hygiene, inappropriate holding equipment, and improper processing techniques are the main reasons for contamination and food poisoning (Brown et al., 2010).

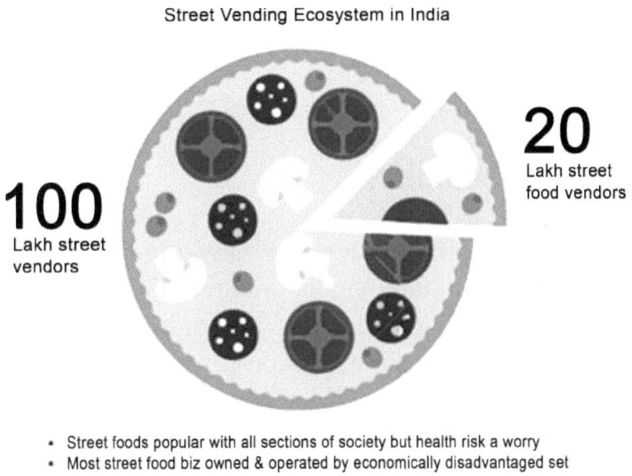

FIGURE 9.1 Street vending ecosystem in India.

Source: Adopted from FSSAI, 2018.

The street food vendors are broadly categorized as (Chakravarty and Canet, 1996):

- A street food vendor who prepares food at home or a rented place and brings it to the stall to sell.
- A vendor who prepares fresh at the counter and sells fresh at the counter; however, some of the mise-en-place is previously done.
- A vendor who prepares food in a small factory or with the help of machines; they work like a cottage industry and then food is sold at the counter.

There is increasing demand for street food in India, so is for vendors. However, there is no significant research that talks about the nutritional aspects of street food in India. A general perception is that the Indian street food is spicy and oily. In general, the energy intake through street food in adults is from 13% to 50% and in children it ranges from 13% to 40%. Although there can be other factors which may influence the energy intake but even if we look at the lower scale, then also it looks like a significant number; therefore, it is important to research more on the dietary and nutritional aspects of street foods in their geographical origin. A study on the nutritional aspect shows that street food are rich in iron and vitamin A but lack vitamin C. Numerous studies have shown concern on increasing obesity, hypertension, and tiredness because of the higher content of fat, trans fat, salt, and sugar in majority of the street foods (Steyn and Labadarios, 2011).

9.2 GAINING POPULARITY

Street food vendors migrate to bigger cities in India from smaller towns. Some of them come for employment opportunities and some come along as dependents, there can be others factors too; however, these two factors remain as main factors for the migration from rural India to urban cities. While looking at the profile of vendors across India, they belong to economically lower section and are normally migrants operating stall for their livelihoods. Although street food vending is not a new profession and is an important part of Indian street vending activity, starting food stall is not labor intensive and is very economical, so, it gives an opportunity for such people to start something own their own. But the vendors also face livelihood challenges. Being an informal system, they are prone to harassment from government authorities. There is no fixed space for these vendors. They are mobile so the fear of losing their livelihood is always on them. To protect themselves, they have to take care of the external factors that may impact their business survival. The system of street food has grown so big that it has become an integral part of urban informal economy and street food has a very strong social and cultural dimension to it. The food generally belongs to a particular community or society; therefore, it brings the cultural and social connect to it and because of the traditional local taste, the street food gets easily accepted.

9.3 ISSUES AND CHALLENGES

Street food is enjoyed for its unique taste, local flavor, low price, and maintaining the traditional touch. It ensures the livelihood of one segment and caters to the need of another. A price-sensitive segment (Fig. 9.2) that eats our on regular basis but is not able to afford restaurant meals, for such customers, street food comes handy. Literature has shown hygiene as one of the biggest concerns amongst consumers but such price-sensitive customers are really not concerned with hygiene, setup, location, or service style. Their main focus is pricing and flavor (Mensah et al., 2002).

Generally, street food carts are not appropriately designed. Most of the time, even the basic facilities like water, napkin, seating, etc. are also not in place. They do have provision for portable water but water for washing of utensils and hands is poorly managed. Sometimes, same water is used multiple times for washing of hands or rinsing of glasses, sometimes the used water is left on the street, and there is no sense of basic hygiene. Most of the vendors do provide dustbins but they get emptied in the nearby places. As result stray animal, insects, and rodents get attracted making the stall unhygienic. Generally, street vendors do have access to washrooms but in places where there is no access to toilets, vendors get forced to excrete in open. After which, depending on the location and availability of water, the hands are washed or not. There was a time when newspapers and leaves were used to serve food but today especially in the tier I cities, vendors have started offering plastic plates, glasses, and cutlery. However, still, in majority of the places, newspaper and leaves are frequently used. The use of newspapers and plants by vendors increase chances of bacterial infection (Mensah et. al., 2002). Vendors with no formal education or food certification have no idea about planning smooth operations and merits of maintaining good personal and space hygiene, insufficient knowledge, lack of equipments, improper processing of food, and irregular storage also results in food contamination (Dawson and Canet, 1991).

FIGURE 9.2 *Kachori*: Popular Indian street food.

Food hygiene and the fear of foodborne diseases remains the biggest challenge to attract customers. The severity of the infection will vary, ranging from mild abdomen problems to life-threatening infections. Mainly these infections are caused by biological microorganisms such as bacteria, viruses, and fungi (Edema et al., 2005). Some of the researches have also found pathogens in the food system especially of the developing countries. Such pathogens may seed short- and long-term diseases like tuberculosis, gastroenteritis, or serious food poisoning. Bacillus cereus is another pathogen which is commonly found in poultry in India. This makes chicken-based street food like chicken fried rice, chicken noodles, and chicken momos more prone to getting infected (Sudershan et al., 2009). Distance from the place of preparing to the stall and the location of the stall has a major role in the amount of bacterial growth. Stalls closer to the kitchen with low human movement were found to have low bacterial count than stalls placed closer to main roads with heavy traffic and human activities (Tambekar et al., 2008). Hygiene and hygiene-related aspects remain the biggest concern of the authorities monitoring and controlling street food vendors. A large number of main dishes were found to be contaminated in random sampling with varying degrees of bacteria's present in the food. The research in various parts of India showed the presence of bacteria *E. coli* in milk and various milk-based products (Singh and Prakash, 2008). Some of the pathogens found in the commonly consumed street food include: *samosas* 25% *Pseudomonas aeruginosa* and 32% *E. coli*, k*achori* 27% *Staphylococcus Aureus*, salmonella spp 36% in *samosa* and *batatawada* and *pakora* 7%. The presence of such bacteria are mainly due to untreated or infected water used for preparation, poor level of washing facilities both for hands and utensils (Tambekar et al., 2008). The presence of Staphylococcus Aureus is mainly due to poor personal hygiene like nails not trimmed, hands running on hair, hands not washed properly, infected clothing, or it may be due to open wounds, cuts, blisters, or dirty scalp (Ghosh et al., 2007). In various parts of India food samples from street vendors were found contaminated with one or more bacteria's. Food items like fried *amritsarumacchi, gulab jamun, athirasai* were found infected with vibrio spp and *E. coli*. The high amount of coliform found in the food is mainly due to mishandling during the transportation. Huices sold in the juice centers were found to have *E. coli* and water had traces of fecal contamination which is used in juices. Random sampling of items like *bhelpuri, sevpuri, panipoori, banta* were found to be contaminated with bacteria (Sudershan et al., 2009).

9.4 HACCP: NEED, ROLE, AND ADOPTION

The street food in India is gaining popularity and is growing at a decent pace. Despite of all the odds, it still attracts price conscious customers on daily or weekly basis. There is still scope of improvement but basic hygiene remains one of the biggest hindrances for not been able to attract customers from the higher segment. The growing concerns from customers and media coverage on foodborne issues have resulted in authorities planning and implementing hygiene standards. One of such programs is HACCP, the program was defined by Codex Alimentarius Commission (CAC) which is globally accepted and is widely followed in star category hotels. The program has seven steps in analyzing the hazards and identifying the critical control points. Having system like HACCP will improve the hygiene standards amongst the Indian street food vendors and will make it safer to be consumed. The HACCP program has seven steps:

- Conduct a hazard analysis
- Identify the critical control points
- Establish critical limits
- Monitor CCP
- Establish corrective action
- Verification
- Recordkeeping

The program starts with conducting and analysis with listing all the steps involved, right from purchasing to getting it served, while looking at the steps, a plan will be prepared which will identify significant hazard and steps to prevent them. The identifying of the critical control point is where the action will be applied to either prevent or eliminate the potential danger. This is followed by setting up critical limits in terms of minimum acceptable norms before it turns into a potential danger. Aspects like PH level of water, time of food exposure, temperature of storage and transportation, etc. are fixed and followed accordingly. Once the critical control points are identified then monitoring of the steps and procedures to monitor are setup. For any issues identified, corrective actions are put in place they are then verified and checked. If all the actions are working well, finally all the recordkeeping is put in place to be used on regular basis and for future reference.

Looking at the HACCP plan, it looks like a full proof system that will help improve the conditions of street food vendors significantly. But the biggest challenge will be training of the vendors and its implementation. Most of the vendors are either uneducated of have only basic level of knowledge; teaching and making them understand the entire system will be difficult. The vendors do understand the importance of maintaining hygiene and are putting their efforts to improve on the hygiene standards; however, formally documenting and monitoring it will be difficult. Government initiatives like *Swachh Bharat Abhiyan* were able to create awareness on importance of keeping the surrounding clean; it had an impact also but there is still a long way to go.

FIGURE 9.3 *Momos*: Popular Indian street food.

The HACCP principles were studied in two of the commonly used street foods in Delhi. *Momos* and *kulfi*; *momos* (Fig. 9.3) is a refined flour dumpling stuffed with finely chopped vegetable or meat and steamed, *kulfi* (Fig. 9.4) is a popular milk-based frozen dessert commonly sold in Indian streets (Sachdev and Mathur, 2017). Results show the need for basic level professional and personal hygiene. Hazards and critical control were identified and preventive actions were suggested. The stuffing for *momos* was raw; it should either be sautéed or gloves or clean hands should be used while handling. The food items and utensils should not be kept on floor, which should be avoided and floor should be sanitized before and after the use. Footwears to be avoided in the area of preparation and hands should be washed and cleaned while handling the dough. The dumplings were steamed for about 10 minutes, which kills the pathogens but does not completely destroy bacteria. Similarly, for *kulfi* boiled milk kills most of the bacteria but adding of *Khoya* (condensed milk) form external supplier increases the chances of cross contamination. The *Kulfi* moulds are not treated or sanitized in any form and adding of the bamboo sticks increases the changes of bacterial growth. The used cones and bamboo sticks were stored in unhygienic condition, which may result is bacterial growth. Similarly, vendors need to be trained on identifying any potential risk and steps to eliminate them. This will not be possible without training and teaching of street vendors. Although, there are licenses to start vending, stringent rules need to be maintained. There can be an exam or certification, post which the license should be issued to all the vendors. It could be made mandatory to have HACCP certification before getting the license and HACCP certificate to be displayed on their respective counters. Government authorities need to be more vigilant and proper action should be taken against the violators.

FIGURE 9.4 *Kulfi* with strawberry syrup.

9.5 CLEAN STREET FOOD HUB

Looking at the challenges and issues faced by the customers and realizing the opportunities for the street food vendors, FSSAI has introduced clean street food hub campaign in India. The objective of the campaign is to create a healthy atmosphere amongst the street food vendors and gain confidence of the customers. Based on the principles of infrastructure, safety, encouragement and promotion, the program will help revive the traditional food items and create awareness for safe and hygienic food (Fig. 9.5) to be sold in Indian streets.

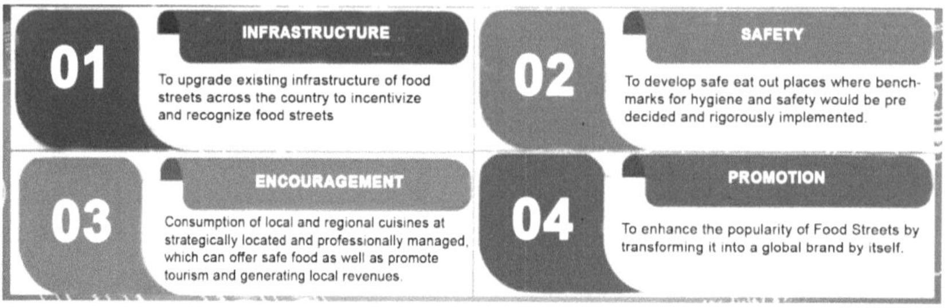

FIGURE 9.5 Food hub campaign.
Source: FSSAI, 2019.

In accordance to the safety regulations, FSSAI has formulated standards and benchmarks to upgrade the standards of the existing street food vendors across the country. The guidelines have developed a Food Safety and Management System (FSMS) which will gradually be implemented pan India. The FSMS program will mainly focus on:

- Location of the stalls
- Existing facilities and required facilities of the vendors
- Details on cleaning and sanitation
- Quality of water and its source of supply
- Waste collection and its disposal
- Any other requirement based on identifying of critical control points to ensure food safety and hygiene.

The plan has already been successfully implemented in Kankaria lake area of Ahmadabad. With around 66 vendors selling different street foods it became India's first clean street food hub. A display of FSSAI certification upfront has helped gain confidence of 12 million people who grabbed a bite on these stalls every year (Dasgupta, 2018). Similarly, Municipal Corporation of Gurugram, Delhi NCR, decided to revamp the food vendors in Gurugram (Fig. 9.6). The vendors will be provided with standard kiosks with nominal rentals. As a pilot run, the government has started the project in Sector 56, Gurugram. Once successful, it will then be replicated in other areas of the city. Along with kiosks, the government will also arrange for drinking water and public toilets The idea is to change the way street food vendors operate making it safe and hygienic for general public (Times of India, 2019).

FIGURE 9.6 Government initiative to transform the street food vending carts.

Implementing of such initiatives successfully will ensure health and hygiene of street food and will ensure clean food for all the consumers. It will also help uplift the socioeconomic status of street food vendors (Fig. 9.7). It will attract more consumers which gradually will help transform the street food image globally (Fig. 9.8). This will not only give confidence amongst consumers but will also help boost the economy. Digitization initiative and acceptance of plastic money even for the smallest of transactions has helped increase sale at the basic level, and on a bigger picture, it will help monitor all transactions taking place in unorganized street food retail market. After receiving several complaints from consumers, FSSAI has issued notifications to all the food business operators, resulting in business operators like Swiggy and Zomato collaborating with only FSSAI certified outlets. In line with the expectations of the consumers, FSSAI is redefining its role as an organization. All the state governments, testing labs, food scientists, and food businesses will be aligned together to achieve common goal, says FSSAI CEO Pawan Agarwal (Economic Times, 2018).

FIGURE 9.7 Traditional street food vendor.

FIGURE 9.8 Modern food vending cart.

9.6 EMERGING STREET FOOD IN INDIA

Irrespective of time, the streets of India will always witness some kind of street food being served. The food items may vary from city to city but right from breakfast to dinner, there will always be one or another kind of street food to offer. The growth of Indian street food could be imagined by the formation of National Association of Street Vendors of India (NASVI), an association which is spread over 26 Indian states and have more than 8,67,426 vendors across. Along with NASVI, FSSAI also organizes National Street Food Festival in Delhi where vendors from all over India participate every year. In general, both males and females operate carts across country, while females prepare more traditional and authentic foods, males are more involved in modern fusion street food that is gaining popularity rather faster.

The traditional street food typically has Indian origin and all the ingredients are locally procured; however, with global fast food and international cuisines gaining popularity, Indian street food vendors are adopting trends and improvising their products, for instance, *noodle, fried rice, momos,* burger, pizza, shawarma were never an integral part of Indian cuisine (Fig. 9.9) but in today's scenario street food without these items are unimaginable. The average purchase of street vendors range from 5 kg to 40 kg per day, although it depends on the food items on offer. Stalls with Indian food items have relatively higher purchase in terms of volume as compared to vendors who offer cosmopolitan food on the other hand. In terms of the value of purchases, vendors offering fusion food has higher value against their Indian counterparts. The purchases range from Rs. 200 to Rs. 2000. Generally, there is not much of a difference in pricing; however, there is marginal difference with the locality of the stalls and some competitive pricing is visible amongst the vendors themselves. The profit margin of the Indian food items is marginally higher compared to mixed street food offerings. Typically, the stalls operate from 10:00 to midnight but vendors offering breakfast start early and who offer late night snacks are open till late at night but this also varies from city to city. Cities where people are awake till late night for office work or travelling have seen vendors operating late night and in rest of the places they close as the city closes.

FIGURE 9.9 Street vendors getting their stalls ready.

The street food is not only limited to streets but is also becoming common in five star hotels, looking at the gaining popularity. Five star hotels host street food festivals on regular basis. The tickets of Indian Street Food Festival, 2018 shows that various segments are getting involved in making it a huge success (Fig. 9.10), from cafes to five star hotels to government bodies all are getting involved.

FIGURE 9.10 Indian street food festivals 2018.

The street food festivals of India offer a variety of delectable food items from across India. The best and the most popular dishes from across states are selected to be a part of the menu during the food promotion. The festivals recreate the street decor and ambiance with carts, street lights, rickshaw, etc. and maintain the authentic taste of the food. Typically, during the festivals one would see *chole khulce* from Amritsar, *chole bhature* from Delhi, *kebabs* from Lucknow, *biryanis* from Hyderabad, *pav bhaji/bhelpuri* from Mumbai, *dosa/Idli* with *sambhar* from southern part of India, assorted lip smacking *chaats* from Mathura and desserts like *misti doi, rosogulla* from West Bengal (Fig. 9.11).

FIGURE 9.11 Favorite Indian Snacks.

All three meals are served as a part of street food but there are some items which are popular and can be eaten in any time of the day; for instance, *idli smabhar, chole bhature, pav bhaji* are some of the famous and popular street foods from India that are eaten as breakfast, lunch, evening snacks, or even dinner. Similarly, different states have different eating preferences (Fig. 9.12).

FIGURE 9.12 Popular dishes during street food festivals 2018.

Source: Times of India, 2017.

Looking at the popularity of Indian street food, a unique concept of "Food Walk" has been started in New Delhi; started in 2011, more popularly known as Delhi Food Walk (Fig. 9.13). Old Delhi streets (Fig. 9.14) are a paradise on Earth for street food lovers. Every lane leads to many by-lanes and all are full with extraordinary street food offerings. The Old Delhi's food is a gastronomic glory that reflects Indian tradition, history, and passion. These food offerings have shaped over many decades and are a witness to Indian history. Delhi Food Walks have numerous options available to choose from. Depending on individuals' preference, one could select a package and cherish accordingly. Some of the popular food walks include Old Delhi food walk, Street food on your plate, Breakfast Trail, Brunch food walk in old Delhi, Mum's kitchen, Nizamuddin, Tea trail, kebab trail, and Gurudwara walk.

FIGURE 9.13 Delhi food walks.

FIGURE 9.14 Preparations in Old Delhi streets.

9.7 GLOBAL INDIAN STREET FOOD

The Indian food was already popular amongst global consumers. Some of the street food has already been introduced internationally but the Indian street food got its global spotlight when it got a chance to be featured in *World Street Congress* in Singapore. NASVI and Singapore Tourism decided to host four of the Indian vendors to showcase their regional street food specialities. Two of them were from Karnataka, one from Delhi, and one from Bihar. The south Indian street food on display was *Chicken 65*, with steamed tamarind and lemon rice, from north of India, it was *Litti Choka* with garlic soaked mutton rice, and from Delhi it was time for *Bhel Puri* with potato crisps flavored with amchoor. Participation like these will help preserve traditional Indian street food. It will make it more professional and will gain confidence amongst international consumers. Such activities will also result in creating more avenues to popularize Indian street food internationally (NASVI, 2019). Indian food was always popular in the United Kingdom. The Indian diasporas present in varies counties of the United Kingdom made it a global cuisine. Since 2000, the Indian dishes are commonly seen in the United Kingdom restaurant segment to an extend where in 2001 British even claimed *Chicken Tikka Masala* as their national dish. Not only the main dishes but also the introduction of Indian street food in the super marts like ASDA, TESCO, and Waitrose shows the emergence of Indian street food at the global level.

Inspired from the name chaat and its flavors, Ashutosh Dua and Shiraz Noor introduced a series of yogurt-based healthy afternoon snacks in the United States. Manhattan-based businessmen believe that, while the street food is not very common in the United States, the ingredients used in the *chaat* are. They further added that Americans are increasingly replacing meals with snacks and most of the Indian food has already been introduced in the United States but *chaat*-based snacks are yet to be introduced. Indian *curries, naan, tandoori* snacks are already quite popular amongst the locals; therefore, they are familiar with the ingredients and introducing Indian street food-based snacks will be acceptable and become popular effortlessly (Cornall, 2016). Similarly in the United Kingdom, two of the Indian origin food and beverage professionals, Sujay Sohani and Subodh Joshi started selling vada pao in London. They started with a small rented space in London and now own two of their outlets with the name of *Shree Krishna Vada Pao* in Hounslow and Harrow (Shreekrishna.com).

9.8 CONCLUSION

Street food in general is economical, has cultural connect, and is an integral part of Indian tradition. The style of operation differs from vendor to vendor but largely only assembly is done at the stall and most of the items are preprepared. The vendors do not have enough space and time to prepare the food at stall; therefore, most of the items are preprepared and assembled at the counter. Some of the vendors purchase ready to serve food items and serve at the stall. Some prepare them in the rented spaces away from the vending machine or stall and transport them every day to the stall. Unlike restaurants and hotels, street food does not hold on to a lot of inventory, thereby, saving on inventory cost. Street food plays a major role in income generation, it indeed has become a great opportunity in India for uneducated or semi educated people to start something on their own. There is no specific or advanced skill required for starting the counter; so, people with little or no skills can also try their luck and start street food vending machine. Traditionally, in India, only males used to operate street food stalls, nowadays, even women are equally seen selling food items in streets. Women sell more of traditional and regional street specialities, and men try their hand in fusion street food like Indo-Chinese or Indo-American or any other variation of fusion street food.

The other side of the street food market is not as rosy as it appears to be. Unfortunately street food has some negative aspects too. The basic hygiene levels have improved reasonably but the conditions are still categorized under moderate to poor. There are efforts being taken by the government but lack of supervision and poor planning has made the progress rather slow. Improving the conditions will help gain confidence of local people and will result in increased footfalls and the impact can directly be noticed on sales. For customers, one of the main reasons to step out for street food is to replace meals at home. They have a great option for evening time snacks and are very economical. The list of the items that are sold the most include *gol gappe, samosa, chole bhature, vada pav, idli, dosa, chicken tikka, momos, roll, and kebabs*.

Looking at the gaining popularity, authorities should plan for more constructive ideas and should be implemented religiously. There are plans in place which have been executed in parts of the country. These plans have actually transformed the working conditions but still there is a lot of scope for improvement. The problem in setting up the place and running it more hygienically needs basic level of training and education. Most of the vendors are not educated, thereby making it even more difficult to execute the plan. Moreover, the customers are also at fault for not being able to keep the surroundings clean. There are places with proper dust-bins being kept but littering can still be seen just outside the bins and all over the place. Planning and implementing it correctly are two different aspects. Government authorities are doing their part by providing the basic infrastructure; simultaneously, vendors and customers need to do their bit. Street foods in India are much more than just food; they reflect culture, tradition of the state that they belong to. Street food was an integral part of Indian society and it continues to remain so; however, over the period of time their service style has changed significantly and the basic infrastructure has also changed for betterment.

KEYWORDS

- **Indian street food**
- **food hub**
- **emerging food trends**
- **street food festival**
- **food promotion**
- **Delhi food walks**

REFERENCES

Brown, A.; Lyons, M.; Dankoco, I. Street Traders and the Emerging Spaces for Urban Voice and Citizenship in African Cities. *J. Urban Studies* **2010,** *47*, 3.

Chakravarty, I; Canet, C. Street Foods in Calcutta. Rome: Agriculture and Consumer Protection Division, FAO, 1996.

Cornall, J. Chaat Takes Indian Street Food to US Dairy Aisle. [Online] 2019. https://www.dairyreporter.com/Article/2016/06/10/Chaat-takes-Indian-street-food-to-US-dairy-aisle (accessed March 7th, 2019).

Dasgupta, I. How FSSAI is Ensuring that Food Offered by Street Vendors and Online Aggregators Meet Safety Norms. [Online] 2019. https://economictimes.indiatimes.com/industry/cons-products/food/how-fssai-is-ensuring-that-food-offered-by-street-vendors-and-online-aggregators-meet-safety-norms/articleshow/66101436.cms?from=mdr (accessed March 5th, 2019).

Dawson, R. J.; Canet, C. International Activities in Street Foods. *Food Control* **1991,** *2*.

Economic Times. [Online] 2019. https://economictimes.indiatimes.com/industry/cons-products/food/how-fssai-is-ensuring-that-food-offered-by-street-vendors-and-online-aggregators-meet-safety-norms/articleshow/66101436.cms?from=mdr (accessed March 5th, 2019)

Edema, M. O.; Omemu, A. M.; Bankole, M. O. Microbiological Safety and Quality of Ready-to-Eat Foods in Nigeria. In *Abstract of the 29th Annual Conference & General*

Meeting on Microbes as Agents of Sustainable Development, organized by Nigerian Society for Microbiology (NSM); University of Agriculture: Abeokuta, 2005.

FSSAI. [Online] 2019. https://fssai.gov.in/foodculture/cleanstreetfoodhub.jsp (accessed March 3rd, 2019).

Ghosh, M.; Wahi, S.; Ganguli, K. M. Prevalence of Enterotoxigenic Staphylococcus Aureus and Shigella spp. in Some Raw Street Vended Indian Foods. *Int. J. Environ. Health Res.* **2007,** *17*, 2.

Janie, P.; Marie, T. Street Food—A Hygienic Perspective. ACM Press: New York, 2010.

Mensah, P.; Yeboah-Manu, D.; Owusu-Darko, K.; Ablorde, A. Street Foods in Accra, Ghana: How Safe are they? *Bull. World Health Organ.* **2002,** *80*, 7.

Muleta; Ashenafi. Salmonella, Shigella and Growth potential of other foodborne pathogens. *East Afr. Med. J.* **2001,** *78,* 11.

NASV. [Online] 2019. http://nasvinet.org/newsite/indian-street-food-vendors-to-get-global-spotlight-at-world-street-food-congress-in-singapore/ (accessed March 5th, 2019).

Shreekrishna.com. [Online] 2019. https://shreekrishnavadapav.co.uk/ (accessed March 15th, 2019).

Steyn, P; Labadarios, D. Street Foods and Fast Foods: How much do South Africans of Different Ethnic Groups Consume? *Ethn. Dis.* **2011,** 21.

Sudershan, R. V.; Rao, P.; Polasa, K. Food Safety Research in India: A Review. *Asian J. Food Agro-Ind.* **2009,** *2*, 03.

Tambekar, D. H.; Jaishwal, V. J.; Dhanorkar, D. V.; Gullhane, P. B.; Dudhane, M. N. Identification of microbiological hazards and safety of ready-to-eat food vended in streets of Amravati City, India. *J. Appl. Biosci.* **2008,** *7*.

Times of India. [Online] 2019. https://timesofindia.indiatimes.com/city/gurgaon/street-vendor.s-to-be-given-carts-space-in-pilot-project/articleshow/68758544.cms (accessed May 3rd, 2019).

CHAPTER 10

Kitchen Design Trends

ABSTRACT

There are numerous factors in the success of restaurant business, of all; repeat business remains the key to success. Hoteliers and restauranteurs are increasingly looking for ways and means to create better experiences amongst customers; in a food and beverage business, food is an integral part to create memories that helps retain customers. To get good quality food; consistently, human resource is an important aspect but for maintaining consistency; planning and designing of the kitchen is equally important if not more. Most often than not, the ownership of restaurants pay very little emphasis on the planning and designing of the kitchens, mostly considered as the back of the house, it gets less attention both; financially as well as aesthetically. With the modernization of techniques and use of technology in kitchen, the level of service has moved higher, so has the customer expectations. The face of kitchen is changing more than ever, with the rise in food activity, the kitchen and kitchen trends have changed, kitchen has moved from the back of the house to become the face of the restaurant. Technology, accuracy, cost effectiveness and customer's interest in food integration has made the change possible.

Kitchen designs are generally based on the space availability, flow of work and convenience of the staff. An ergonomically planned kitchen provides a positive work-flow which helps in efficient food production leading to profitability. There are few basic points that need to be accommodated while getting the blue-print for construction of a kitchen; provision of drainage, air-circulation, non-slippery floors, no intersection between kitchen and service staff etc. are some of them. The basics of planning remains same but changing trends have added more aspects to be considered while planning a modern day kitchen. One such aspect is sustainable kitchen; sustainable kitchen need to have aspects like efficient waste management, proper tools and measures for energy conservation, recycling of waste and water and minimizing food wastage. In the chapter we will look at the transformation of planning considerations from traditional planning to future trends and discuss the latest kitchen trends in India with special emphasis on sustainable kitchen trends.

10.1 INTRODUCTION

When it comes to kitchen planning, designing and restaurant setup, there are three key stakeholders; customers, employees, and the owners. Although, all three of them work toward the same goal but their expectations and demand differ significantly. Customers are always looking for new and better experiences, employee prefer operationally more suitable and feasible kitchens and owners focus on investment and the return on investment. Together it means a kitchen needs to have a balance of functionality, aesthetics, and at the same time should be able to meet the need of the customers and the owners. Generally, in star category hotels and specialty restaurants aesthetics dominate without compromising on the operational functionality. However, in majority of food outlets only functionality is looked at with maximum utilization of space. Irrespective of the design, the kitchen planner should be able to maintain the balance of safety, efficiency, and cost effectiveness. Kitchens are very complex spaces but a well-designed kitchen will have maximum efficiency in minimum space. The output will depend on the product movement in the kitchen; the flow of work needs to be unidirectional with either no cross sectioning or minimum crossing paths. There has to be sufficient space for the movement of the kitchen employees but not too much that they have to cover larger distances to get the food prepared. The induction of modern technology in the kitchen has made the work flow much smoother and has also resulted in more safer and efficient kitchens (Pehkonen et. al., 2009).

There are different types of kitchen layouts but in the modern day kitchen planning, the layouts can broadly be categorized under two categories; open and closed kitchens. While selecting the design, work triangle becomes the foundation of planning. In kitchen mostly the movement is between the areas of food storage, food preparation, and cooking; hence, classified as kitchen movement triangle. However, movement around work table and pot wash is also significant but varies from outlet to outlet.

The most common layout seen in kitchen includes straight line, L-shaped, U-shaped, G-shaped, and galley. The basic idea to have a designed kitchen is to place the equipment, food, furniture, and human resource in a way that there is optimum utilization of available space with least employee movement (Llewellyn, 2004).

Moreover, incorporation of latest food trends like molecular gastronomy, farm to fork, sustainable practices, increased awareness of preserving natural resources have resulted in changing the design, shapes, and patterns of kitchens. With globalization, the distances are shrinking, more and more people are getting exposure of international cuisines and cultures, all such factors are pressurizing chefs and managers to introduce new food products, cuisines, and trends. These inclinations give a new dimension to planning and designing of kitchens. One of such trends is open kitchen. It started gaining popularity in 1990s and became so popular that today almost all the restaurants small or big, stand-alone or of a star category hotel have open kitchens (Baraban and Durocher, 2010).

There is a drastic change in the deliveries and procedures of the food industry, which is leading to many challenges. Today's kitchen force is working on genetically modified foods, use of local and fresh raw materials to provide the most attractive and healthy food to the guests. Not only the use of organic ingredients but chefs and managers are also trying to implement energy and water conservation strategies and molding the green future of the food industry (Hensley and Donohue, 2008). Expanding kitchen scope has also increased the demand to understand the importance of space allocation with the study of space optimization (Ghiselli et al., 1998). Increasing competition and the effort to create new experiences for customers; the menus, products, procedures, equipment's area is also getting complex with increasing complexity; providing safer environment for employees and customers is also becoming challenging. Sometimes the management gets too involved in creating new aspects, the basic level of expectations like health, safety, and service attention gets compromised. Food safety is the prime concern for any restauranteurs and its safety lies in the way food is handled in kitchen. Kitchen planning and its design will be crucial in food handling; one of the biggest challenges that is faced by the chefs is the decreasing size of kitchens in such situations. Employees have to compromise with the safety and quality of food in smaller kitchens. The basic rule of employees crossing the path gets disobeyed, which may lead to compromising on employee and food safety (Stipanuk, 2006).

10.2 ROLE OF KITCHEN DESIGNS

There are evidences where the role of restaurant design with an open kitchen has positively impacted the bottom line. It helps in creating a positive image in customers mind. The perception and overall experience of the customers is elevated which results in repeat business. One of the biggest differentiator for hospitality industry is its tangibility but non-tangible aspects like aesthetics, surroundings, décor, and ambiance add to the overall experience and create a sense of the quality of service that customers may expect to receive. The design and décor not only helps in creating perception but it also impacts the overall satisfaction of customers. The aesthetics, design, and décor not only impacts customers experience but it also adds value to employees working in the restaurant (Ransley and Ingram, 2001). Having a well-designed kitchen increases the efficiency of the kitchen staff, takes lesser time, is safer and the overall productivity of the kitchen operation increases (Fig. 10.1). The use of technology in kitchens, placement of equipment, and a well-trained staff has direct impact on efficiency and productivity; however, in smaller setups introduction of bigger machines and tools becomes difficult. A well-planned kitchen reduces stress level of the employees, which keeps them fresh and motivated to perform better. Therefore, planning and designing has direct impact on functionality and efficiency of kitchens; it maximizes the productivity, improves quality, and saves financially and keeps employees motivated.

FIGURE 10.1 Example of a balanced kitchen.

10.3 SUSTAINABLE KITCHENS

The term sustainability not only refers to protecting natural resources but is also multidimensional. Having a sustainable kitchen refers to the use of eco-friendly material and should be able to sustain the day-to-day food and beverage operations effectively. The importance of backward integration will help build a sustainable kitchen. However, the biggest hindrance to build such kitchens is the availability of funds and the return on investment. Mostly, business look to breakeven as soon as they could but introduction of eco-friendly systems are very expensive. As a result the breakpoint time increases and it discourages restaurateurs to build a complete sustainable kitchen design. The design to have all eco-friendly equipment helps in energy conservation but it costs lot more than regular kitchen equipment. The planning and designing of commercial kitchens has various components; there is a long list to be considered. However, instead of building a complete eco-friendly kitchen, chefs are increasingly looking to replace equipment that consumes low energy and has less impact on the environment. Therefore, the star rating that defines the energy consumption becomes relevant while making the purchase decision.

The main aspects that need to be kept in mind while planning and designing a sustainable kitchen include:

- Fittings and fixtures
- Floorings, walls, and ceiling
- Space planning
- Fire safety
- Hygienically prepared food
- Water supply and drainage
- Ventilation
- Lighting
- Storage
- Pest control
- Waste management

While it may not be financially feasible to plan all the aspects on sustainability but depending on the size, budget, and need of the outlet, some of the major components could be looked at to design the kitchen as eco-friendly as possible.

10.4 FITTINGS AND FIXTURES

While planning about the sustainable kitchen, we need to start thinking about all the material that will be used in the kitchen. All the aspects from building of the kitchen to running the kitchen for day-to-day operations should be considered. The chefs, managers, and decision makers need to reflect on the product usage, its impact on the environment and society. The source of the raw material needs to be looked at; its production cycle to be studied and environmental impact that it may have on its production becomes crucial. Similar to the farm to fork cycle of food, the entire cycle of material used; from its production to disposal, needs to be well researched and considered while making a design. One of the simpler approaches is to keep all the items to minimum possible, only the equipment and utensils which are extremely important should be planned and purchased. Keeping the kitchen spaces smaller will help but it should not get too clumsy for the employees to work on daily routine. Items made with the recycled material should be encouraged and reusing the items available within will be financially helpful and will also protect the environment by reducing the carbon footprints.

10.5 NATURAL RESOURCES PLANNING

The biggest issue that hospitality industry is facing is the optimum utilization of the available natural resources. The kitchen appliances with highest rating of energy savings should be encourages to be used in hotels and restaurants. It will save energy, protect the environment, and reduce on the electricity expense. Air conditioning, provision for fresh air, refrigeration, and dish washers consume the maximum amount of energy, so manufacturers with latest technologies and least energy consumption should be considered while purchasing these items. Electricity consumption has reduced significantly because of the use of LED bulbs and motion sensors.

10.6 FOOD WASTE DISPOSAL

Food waste and its proper disposal has become a big issue in all the major cities of India. Mostly, the food waste is dumped in open area that results in formation of methane gas and is considered more dangerous for the environment than carbon dioxide. The landfill in Delhi NCR and in similar cities are increasing by the day and creating health issues to infants and elderly people. The government has taken steps to reduce the waste from reaching the landfills to as much as possible but still the problem persists. Some of the hotels have started converting the food waste to compost with their premises. It helps in multiple ways but still majority of the restaurants are getting their waste supplied to the landfills through government organizations. The star category hotels have more space and higher budgets, therefore, they can afford to implement such practices but for small and stand-alone restaurants it becomes difficult to introduce such practice in daily routine; they are and will remain dependent on external bodies for processing their food waste.

FIGURE 10.2 Kitchen waste disposed in open pit and covered with straw to help compost naturally.

10.7 INDIAN HOTELS

The sustainable designs are in trend across hotels and customers are increasingly looking for eco-friendly hotels. According to Economic Times, in India 2900 times the term "green hotels" was searched in Google annually against 60,500 times globally. Such results show the inclination toward green hotels. With the amount of awareness being created around sustainability practices; it will continue to increase with time. Moreover, eco-friendly practices have direct impact on the bottom line. Initially, the installation is expensive but in the longer run it pays off. Financial impacts are there but the businesses are also getting environment friendly and want to involve in practices that are good for protecting environment (Fig. 10.2). The example of Orchid Hotel, Mumbai is a classic example of saving around 3.5 million in a year just by switching to CFL lightings. Similarly, other hotels also are implementing various practices to build greener hotels; initiatives like CFC free air conditioning, use of solar energy, water harvesting, reusing materials, and recycling waste are gaining popularity.

Considering these factors while planning and designing kitchens will lower the cost significantly. It will reduce water consumption and lesser amount of carbon footprints. The procurement of local material for construction reduces cost and helps in reducing carbon foot prints. According to Krishan Singh, business head at Yatra.com while planning and designing sustainable kitchens, the cost of material is 15–30% higher than the regular material used but through sustainable initiatives if implemented diligently, a hotel can save up to 20–25%. The difference looks marginal if compared to the initial cost but the cost for preserving and saving natural resources for future generations is priceless.

10.8 FOOD AND BEVERAGE FRONT END LAYOUT AND DESIGNS

One of the defining features of the continuing evolution of the food and beverage service space in India from unorganized to organized formats has been a focus on how the front end looked to a consumer. The changing sociocultural dynamics, economic growth, exposure to western expression of food and beverage hospitality all have had a tremendous impact on the Indian consumer and their expectations from the organized food and beverage service formats in India. About two decades back it was great to see arrival of air-conditioned restaurants and improvements in the overall hygiene levels being maintained. The process of value addition has continued beyond just temperature control or fly catchers being installed. Technology too has played its role as 'WiFi' enablement has become a norm across almost all organized F&B outlets across formats, and it is common to find businesses using point of sale applications that allow customers to place their orders with ease and pay their bills at the table digitally as well. So much has changed over the last two decades and more is expected in the years to come.

10.9 INTERESTING TRENDS

The F&B outlets of the day are fun, lively, and an expression of what the customer seeks in terms of ambiance, comfort, and convenience. The restaurant owners, companies, or individuals alike recognize that it is the whole meal experience that gets a customer to come back and keep doing that. They recognize that it takes a lot to get a new customer in and takes even more to retain the customers, and a big part of the value to customer comes from the way the place (front end) looks and feels. The food and service combine must be complemented with interiors, temperature control, lighting, and the music to set the stage to create value for the customers. Based on the format, the interiors are a representation of the investment made. The fine dining restaurants go all out to entice the customers with their opulent interior design and luxurious setting to match the expectations of the customers. The casual dining and quick service restaurant formats are more about comfortable seating and interiors that are chic and young. These are also designed to enable self-service or quick service based on the food and beverage concepts that they offer. The bars and pubs are looking to give a design statement through their layout and interiors including the lighting, music, and TV screens and often a visible microbrewery to cap the look. Some of the popular trends that are quite visible are:

- Rustic interiors
- Exposed ceilings as against false ceilings that were popular few years back
- Focus on number of covers that are set up to ensure revenue per square foot is maximized or optimized such that customer experience is not adversely affected
- Energy-efficient lighting with many outlets going for creative fixtures that use recycled materials
- Open kitchens that are like cooking islands set with in the service front end, often increasing the air conditioning requirement as well
- Electronic menus with attractive graphics, tech-driven POS systems, and digital billing systems

- Creative expression, artwork, graphic designs and such on walls, ceilings, etc. to convey the theme, attract a target customer profile
- Restrooms too are game when it comes to creating and communicating a design theme, way different from the old times
- Restaurants with a focus on entire families as target customer segment have designed the service front end inclusive of kids areas/ play areas
- Focus on sustainability that translates to use of energy efficient cooling systems, lights, recycled materials, and even furniture that is reused/recycled
- Use of light weight, moveable furniture such that the dining spaces can be easily reconfigured to fit the need of catered events on as required basis akin to a pop-up restaurant
- How are branded chain formats different?

The international as well as domestic brands in the QSR, café, casual dining, or any other organized segment are quite focused on ensuring that brand standards are adhered to by the respective franchisees. The expected conformity to standard operating procedures (SOPs) covers all aspects of their operations including menus, service processes, staff uniforms, staff training, and even purchases with a sole objective of ensuring that a customer is able to experience consistent quality across any outlet for the brand. How does this expectation get translated when it comes to the look of the place, the interiors, the furnishings and other aesthetic elements? Is each Pizza Hut, TGI Friday looking exactly the same across India or are they different?

The question is not as easily answered as it is raised. It is clear that a strong expectation of brand compliance is pushed for as a result of which the restaurant branding and in some cases the furniture and fixtures across stores (restaurants) are identical. What is quite easy to see is the Indian adjustment to brand standards by the franchisees. This was quite difficult for international brands to understand and accept as they arrived in India for business but soon enough they were dealing with regional request for menu changes, accommodating for design and layout differencing, and in essence understanding that India is all about unity in diversity. In other words they had to redefine variations that were completely unacceptable and thus deal breakers as against those where some latitude had to be allowed. This process was handled by setting up master franchises who were then expected to manage the brand expectations for those that were part of their portfolio. In many cases, this has been a tough act for international brands as their stores got to look different and sell different across India.

10.10 ADVANTAGE OF STAND-ALONES

The stand-alone food and beverage operations, of course, have no limitations as to how they would like to operate; in fact many domestic brands that have grown into chains have allowed their new stores to look different, based on their size, location, and market. For a stand-alone the way the service space looks is often as much an expression of the service level and market segment as it is about the creative expression of the ownership and their personal statement of design. This aspect is also guided by the spirit of innovation and creativity that is the hallmark of entrepreneurship. There are plenty of cafes, fine dining restaurants, and bars that will stand as testimony to the earlier mentioned statement. Investors are delighted with efforts by food and beverage entrepreneurs trying to minimize expensive décor and go for options that make the place look fresh and cost less.

There are loads of F&B consultants, big and small, who are willing to guide, design, implement set ups of kitchens as well as service front ends as a living. These professionals are often folks who have worked in the industry for a few years and based on their experience branched off into the design aspects of the hospitality spaces, including interior design. As the food and beverage segments expand into the tier 2 and 3 cities across India, demand for these professionals is going to increase. They are often assisted by interior designers, architects, and other experts as they put together and implement projects. Here again stand-alones have the advantage of being able to be creative while the branded chains limit their new stores to use existing team of professionals at a (usually exorbitant) price to set them up, in fact even suppliers of furniture and fixtures are restricted.

10.11 SERVICE FRONT AND THE KITCHEN BACK END IN SYNC

As reported in earlier chapters, the focus of the organized food and beverage service formats has been to offer a much superior, professional experience to the demanding Indian customer. This needed immediate attention to the service ambiance—air conditioning, aesthetics, and comfortable seating as well as a lot of focus on the HACCP aspects. Kitchens were to look sharp, clean, and well-designed and ventilated, with stainless steel being used extensively. Trends like open kitchen have taken this to another level, depending on how "open" is interpreted, kitchen spaces are increasingly either visible to customers or even set up right in the middle of the service front end such that the concept of a production back end that is invisible to customers is lost completely.

The trends mentioned earlier as well as the changing service expectations require the service and production systems to be designed such that the operations run seamlessly, here again technology has left a big foot print. Hand written kitchen order tickets (KOTs) are being replaced with LED screen displays giving an update on order completion status, that said kitchen design has seen less of creative flair than the service end but that may soon change as food tech is guiding setting up of virtual restaurants with production ends that are invisible to customers often located in areas where the cost/rentals of real estate is cheap. Strengthening delivery-based models based on virtual restaurants have the ability to redefine and challenge the brick and mortar format itself. It would be interesting how this space plays out in years to come.

10.12 FUTURE TRENDS

- Technological Innovation: Kitchen planning and designing continues to create and adopt latest technology and innovation; this provides ground for all the planners to keep adding latest innovations in their designs and also provides a platform for the technology innovators from the information technology sector. Innovation and adoption of such technologies keeps the competitive advantage, it also helps to cater various needs of the customers and employees. However, financial implications remain the biggest hindrance in incorporating latest technology. The preferences and demands of new age customers are quite different and change more frequently, especially with the millennials; they get fascinated with the use of technologies like robots in the live kitchen but such concepts misses the human touch of the food preparations, which remains the key for a good quality and consistent meals. Understanding such needs and demands is important to develop future trends and will be helpful in becoming self-sustainable and profitable in the longer run.

- Digital Kitchens: The primary function of the kitchen is cooking food but in the times of social media marketing, kitchen is also acting as a median of social interactions and these interactions help build social bonds. But in future, the interaction in kitchen is going to be through digital kitchens, some components of digitization are already seen in hotels but the use of technology will be seen more in the coming future. Having more of cameras around cooking with the live access to customer in the restaurant could be made possible; the KOT printers have largely been substituted with the LCD display and orders are coordinated through the screen in the kitchen.
- Intelligent Kitchen: The display screens in kitchens are already been seen but having an intelligent kitchen will help the chef's replicate standard recipe every time they cook the food. There will be prerecorded recipes and with the help of a camera and internet of things, the cooking process will be displayed. If there is any deviation in the cooking process, an alert or a message will be reflected in the screen (Bonanni and Selker, 2005). This will prevent any deviation from the preset recipes, however, for implementing such techniques, financial consideration and operational possibilities remain biggest challenge.
- Virtual Kitchens: The online business is not only growing in FMCG sector but online business for food is also all time high. Customers are ordering food online and the likes of Swiggy and Zomato picks the food from the respective outlets and gets delivered to the destination. In such cases customers have not been able to look at the kitchen and it works as a virtual kitchen. The gaining popularity of social media and its use as the main tool for marketing has changed the food economics. With the online food posts, customers are increasingly posting the steps behind the main cooking. These steps also cover the food preparation area; hence, the kitchen pictures get posted. These pictures of the virtual kitchen help attract customers while placing orders online.
- Smart Open Storage: The storage spaces in kitchens and stores are generally covered from a HACCP point of view. It is important to

follow the norms and keep the space safe from pests but increasing transparency, space constraints, cost control, and guest interactions with the kitchen staff are resulting in more open spaces. As a result smart ways of storage is evolving with time, use of roll out trays, caddies for pots and pans, hanging racks is frequently seen.

- Going Native: More and more people are looking for tradition and authentic food. The demand for local and seasonal food has changed the food service pattern. With the local specialty, customers are also looking for the ambiance, décor, and service style as was done traditionally. Such demands tend to change the designs of the kitchen; however, these demands could be short timed but are crucial in brand building.
- Transparency: Food is increasingly helping in creating the perception about hotel or restaurant brand, increasing number of Michelin star restaurants in star category hotels is a testimony of gaining popularity. With the new age customers, the distance between customers and employees is decreasing, customers are looking for more transparent services and experiences as a result open kitchen is gaining popularity. Open kitchens give a sense of personalized service, the interaction of customer with the chef while preparing the meal adds to the experience.
- Personalized Services: Food is gaining popularity world over. In India, almost every television channel runs a food show as a result there rare increased numbers of celebrity chefs and it also encourages young chefs to look forward to culinary as a preferred career option. In general, such developments are changing the food landscape in multiple ways with food gaining popularity kitchens are also becoming trendy and modular like personalized valet. We will soon see personalized chefs serving in star category hotels, the personalized dining room will have an island kitchen where the guests can see the cooking and can interact with the chefs, it will add to the overall personalized experience of the customers.
- Colors in Kitchens: There is a big contrast in colors been seen in QSR, stand-alone restaurants and star category hotels. The QSR kitchens look more lively and vibrant with the use of bright colors; colors like red, yellow, green, and blue are frequently used. However, the star category hotel prefer to go for metal finishes and to add to the luxury factor, colors like gold, silver, or bronze

are seen more often. The use of monochrome themes with hues of black and white is also gaining popularity.

10.13 CONCLUSION

Hospitality industry continues to be the leading job provider in the world. As per the records of world travel and tourism council, it contributes to 10% of the global GDP and one in every five new job opening is from hospitality sector. It is also one of the fastest growing industries globally with such tremendous growth trajectory; adoption of latest technologies becomes apparent and reducing spaces with increased productivity will be the focus area. Food production is not a revenue generating department, unlike front office and restaurants and is not visited by the customers very often. Therefore, investing in technology and its return on investment becomes prime factor to be considered while adopting new technology. The idea of sustainable kitchen designs have come a long way. It is more than just a buzz word and is religiously been implemented across sectors. A well-planned sustainable kitchen will follow holistic design pattern that enables healthy and safe food consumption patterns and uses the natural resources responsibly. The use of toxic materials is avoided to the best extend possible and the use of nontoxic substances is in fashion. Materials free of formaldehyde and low volatile organic compounds are in use. At the same time, commercial kitchens need to be ergonomically designed. It should be functional, should sustain wear and tear, and meet the customers ever changing demands. Once it is decided to design a sustainable kitchen, there are certain questions that need to be answered. Are there any financial constraints? Is the design complimenting the theme of the restaurant? Is it is practically a feasible design, keeping the work flow in mind? Will it last long? Will it incorporate the fads that keep reoccurring for shorter period of time? The list is endless and depending on the preferences, it will differ from outlet to outlet. However, the consumption of three main elements of Earth, air, water, and sunlight need to be considered while planning to build a suitable kitchen. It should be designed in a way that the natural resources are optimally utilized and minimum damage is done to the environment while using these resources.

There is a lot of pressure from customers and government organizations for adopting environment friendly practices. Therefore, not only the daily purchases, even the planning and designing of future kitchens has to be around green and eco friendly practices. To keep pace with the customers' expectations, it becomes important to be a market leader and set trends for future. Observing current design trends and adopting the circumstances as per the market will be essential to maintain edge over competitors. More use of technology and introduction of gadgets will be visible in future kitchens. Devices like video camera, sensors, microphones, and remotely controlled kitchens will continue to gain popularity. Evolving future trends in luxury dining, sustainability, backward integration, adoption of technology, and endless innovation are constantly changing the future of planning and designing. In such a scenario where trends are changing more than ever, it becomes the prime responsibility of senior chefs, restaurant managers, kitchen planners, designers, and architects to guide the future leaders on acquiring required skills and knowledge to become the innovators of future.

KEYWORDS

- **kitchen designs**
- **modern kitchen**
- **sustainable kitchen**
- **natural compost**
- **future trends**
- **food and beverage layout**

REFERENCES

Baraban, R; Durocher, J. Successful Restaurant Designs, 3rd ed.; Wiley Publications: New York, 2010.

Bonanni, L.; Lee, C. H.; Selker, T. Counter Intelligence: Augmented Reality Kitchen, Extended Abstracts of Computer Human Interaction; ACM Press: New York, 2005.

Kivela, J.; Crotts, J. C. Tourism and Gastronomy: Gastronomy's Influence on how Tourists Experience a Destination. *J. Hosp. Tour. Res.* **2006,** *30*.

Llewellyn, M. Designed by Women and Designing Women: Gender, Planning and the Geographies of the Kitchen in Britain 1917–1946. *Cult. Geogr.* **2004,** *11*(1).

Stipanuk, D. Hospitality Facilities Management and Design, 3rd ed.; American Hotel and Lodging Association, 2006.

Pehkonen, I.; Takala, E. P.; Ketola, R.; Viikari-Juntura, E.; Leino-Arjas, P.; Hopsu, L., Nykyri, E. Evaluation of a Participatory Ergonomic Intervention Process in Kitchen Work. *Appl. Ergon.* **2009,** *40*(1).
Ransley, J.; Ingram, H. "What is 'Good' Hotel Design?". *Facilities* **2001,** *19*(1–2).

Index

D

E

F

H

I

S

T

U